Marcus Rammes

The Top Quark and its Electroweak Couplings

Marcus Rammes

The Top Quark and its Electroweak Couplings

Measurement of the pp → ttγ Inclusive Cross Section in the Semi-Leptonic Decay Channel with the ATLAS Detector

Südwestdeutscher Verlag für Hochschulschriften

Impressum / Imprint

Bibliografische Information der Deutschen Nationalbibliothek: Die Deutsche Nationalbibliothek verzeichnet diese Publikation in der Deutschen Nationalbibliografie; detaillierte bibliografische Daten sind im Internet über http://dnb.d-nb.de abrufbar.
Alle in diesem Buch genannten Marken und Produktnamen unterliegen warenzeichen-, marken- oder patentrechtlichem Schutz bzw. sind Warenzeichen oder eingetragene Warenzeichen der jeweiligen Inhaber. Die Wiedergabe von Marken, Produktnamen, Gebrauchsnamen, Handelsnamen, Warenbezeichnungen u.s.w. in diesem Werk berechtigt auch ohne besondere Kennzeichnung nicht zu der Annahme, dass solche Namen im Sinne der Warenzeichen- und Markenschutzgesetzgebung als frei zu betrachten wären und daher von jedermann benutzt werden dürften.

Bibliographic information published by the Deutsche Nationalbibliothek: The Deutsche Nationalbibliothek lists this publication in the Deutsche Nationalbibliografie; detailed bibliographic data are available in the Internet at http://dnb.d-nb.de.
Any brand names and product names mentioned in this book are subject to trademark, brand or patent protection and are trademarks or registered trademarks of their respective holders. The use of brand names, product names, common names, trade names, product descriptions etc. even without a particular marking in this works is in no way to be construed to mean that such names may be regarded as unrestricted in respect of trademark and brand protection legislation and could thus be used by anyone.

Coverbild / Cover image: www.ingimage.com

Verlag / Publisher:
Südwestdeutscher Verlag für Hochschulschriften
ist ein Imprint der / is a trademark of
OmniScriptum GmbH & Co. KG
Heinrich-Böcking-Str. 6-8, 66121 Saarbrücken, Deutschland / Germany
Email: info@svh-verlag.de

Herstellung: siehe letzte Seite /
Printed at: see last page
ISBN: 978-3-8381-3806-0

Zugl. / Approved by: Siegen, Universität Siegen, Dissertation, 2012

Copyright © 2014 OmniScriptum GmbH & Co. KG
Alle Rechte vorbehalten. / All rights reserved. Saarbrücken 2014

Contents

0	**Preface**	**1**
I	**Introduction**	**3**
1	**The Standard Model of Particle Physics**	**3**
	1.1 Gauge Groups and Symmetries	3
	1.2 Quantum Field Theories (QFT)	3
	1.2.1 Lagrangian and Equation of Motion	4
	1.2.2 Neutral Scalar Field	5
	1.2.3 The Dirac Field	6
	1.3 The Weak Force	8
	1.4 Local Gauge Invariance and Quantum Electrodynamics	9
	1.5 The GSW Theory	11
	1.6 The Higgs Mechanism	13
	1.6.1 Spontaneous Symmetry Breaking	13
	1.6.2 The Higgs Mechanism in the Standard Model	15
	1.6.3 Yukawa Couplings	18
	1.7 Quantumchromodynamics	19
	1.8 Renormalization	20
	1.8.1 QCD and the Renormalization Group	23
	1.8.2 Asymptotic Freedom and Quark Confinement	24
	1.9 The Standard Model	24
	1.10 Quark Mixing	26
	1.11 Challenges to the SM and Possible Extensions	29
	1.11.1 The Hierarchy Problem	29
	1.11.2 CP Violation	30
	1.11.3 Supersymmetry (SUSY)	30
	1.11.4 Kaluza-Klein Excitations	31

2 The Top Quark 32
- 2.1 Properties of the Top Quark 34
  - 2.1.1 Rôle of the Top Quark Mass in EW Precision Physics 34
  - 2.1.2 Electric Charge of the Top Quark 35
- 2.2 Top Quark Physics at Hadron Colliders 37
  - 2.2.1 Parton Model and Parton Distribution Functions (PDFs) 38
- 2.3 Top Quark Production Mechanisms 39
- 2.4 Top Quark Decay 43

3 Electroweak Couplings to the Top Quark 47
- 3.1 Approaches to Anomalous EW Top Quark Couplings 48
- 3.2 Top Quark Pair Production with an Additional Photon 49

4 The Large Hadron Collider 55
- 4.1 Proton Pre-Acceleration 59
- 4.2 LHC Performance 59
  - 4.2.1 Beam Lifetime 61
- 4.3 Luminosity Measurement 62
- 4.4 The LHC Detectors 62
- 4.5 Pile-up 63
  - 4.5.1 Pile-up Suppression 64

5 The ATLAS Detector 66
- 5.1 Coordinate System 66
- 5.2 The Inner Detector 68
  - 5.2.1 The Pixel Detector 69
  - 5.2.2 The Silicon Strip Detector 72
  - 5.2.3 The Transition Radiation Tracker (TRT) 72

5.3 The Calorimetry System 73
  5.3.1 The Electromagnetic Calorimeter 74
  5.3.2 The Hadronic Calorimeter 75
5.4 The Muon Spectrometer 77
  5.4.1 Monitored drift tubes (MDT) 78
  5.4.2 Cathode Strip Chambers (CSC) 78
  5.4.3 Resistive Plate Chambers (RPC) 79
  5.4.4 Thin Gap Chambers (TGC) 79
5.5 The Magnet Systems 80
  5.5.1 The Central Solenoid 80
  5.5.2 The Barrel Toroid 80
  5.5.3 The End-Cap Toroids 81
5.6 The Trigger and Data Acquisition (DAQ) System ... 82
  5.6.1 Level 1 (L1) Trigger 83
  5.6.2 Level 2 (L2) Trigger 85
  5.6.3 Event Filter (EF) Trigger 86
  5.6.4 Data Storage and Processing 86

II Analysis 88

6 Monte Carlo Generators 88
6.1 MC@NLO 88
6.2 POWHEG 89
6.3 ALPGEN 89
6.4 AcerMC 90
6.5 WHIZARD 90
  6.5.1 The Matrix Element Generator O'Mega 90
  6.5.2 Phase Space Integration 91
6.6 Parton Showering and Hadronization 92
  6.6.1 HERWIG 93
  6.6.2 PYTHIA 93

		6.6.3	PHOTOS	94
		6.6.4	JIMMY	94

7 Physical Objects and Reconstruction 95

    7.1 Electrons 95
        7.1.1 Electron Identification 96
        7.1.2 Converted Photons 97
    7.2 Muons 97
    7.3 Jets 99
        7.3.1 Infrared and Collinear Safety 99
        7.3.2 Jet Algorithms 100
        7.3.3 Jet Reconstruction 101
        7.3.4 Jet Energy Calibration 102
        7.3.5 Jet b-Tagging 103
    7.4 Photons 105
        7.4.1 Photon Recovery 106
        7.4.2 Photon Identification 108
    7.5 Missing Transverse Energy 114
        7.5.1 $\not{E}_T$ Calibration and Calibration Schemes 115
    7.6 Isolation 116
        7.6.1 Calorimeter Isolation Corrections 117

8 Dataset 118

    8.1 Data Streams 119
    8.2 Good Run Lists 120

9 Object Definitions and Event Selection 124

    9.1 Electrons 124
    9.2 Muons 125
    9.3 Triggers and Trigger Matching 126
    9.4 Photons 127
        9.4.1 Photon Isolation 128

9.4.2 MC Correction of Photon Shower Variables . . 129
9.5 Energy Rescaling 130
9.6 Jets 132
    9.6.1 b-Tagging 133
    9.6.2 Electron-Jet Overlap Removal 134
    9.6.3 Photon-Jet Overlap Removal 134
9.7 Missing Transverse Energy 135
9.8 Treatment of EMC Hardware Failures 135
9.9 Event Selection 137

10 Signal and Background Modeling   141
10.1 Simulation Chain 142
10.2 Simulation of Signal Events 142
10.3 Background Samples 146
10.4 Signal Phase Space Overlap Removal 148
10.5 Pileup Reweighting 154
10.6 Event Weights 154
    10.6.1 Object Weights 156
    10.6.2 Final Event Weight 157
10.7 MC Uncertainties 158
10.8 Event Yields 159
    10.8.1 Full Event Selection and Signal Efficiency . . . 160

11 Cross Section Measurement   170
11.1 Identification of Hadron Fakes 171
11.2 Choice of the Isolation Discriminator 172
11.3 Modeling of Signal and Background Contributions . . . 173
    11.3.1 Template Fit 175
    11.3.2 Prompt Photon Template 176
    11.3.3 Hadron Fake Template 181
11.4 Estimation of Prompt Photon Background Contributions 183

      11.4.1 Background from Electrons Mis-identified as Photons 186
      11.4.2 W+Jets+γ Background 197
      11.4.3 Multijet Background (QCD+γ) 199
      11.4.4 $t\bar{t}$ Background 206
      11.4.5 Other Background Contributions Estimated from MC Simulation 208
  11.5 Result of the Template Fit 214

12 Systematic Uncertainties   218
  12.1 Method 218
  12.2 Pseudo Data 219
  12.3 Sources of Systematic Uncertainties 220
      12.3.1 Jet Modeling 220
      12.3.2 b-Tagging Performance 227
      12.3.3 $\not{E}_\mathrm{T}$ Uncertainties 227
      12.3.4 Electron, Photon and Muon Performance 228
      12.3.5 LAr Hardware Failure 229
      12.3.6 PDF Uncertainty 229
      12.3.7 $t\bar{t}$ MC Modeling 232
      12.3.8 $t\bar{t}\gamma$ NLO Calculations 236
      12.3.9 Uncertainties from Prompt Photon Background Estimations 236
      12.3.10 Template Modeling 237
      12.3.11 Possible Uncertainties due to Pile-Up Effects . . 238
  12.4 Combination of Uncertainties 238

13 Results   243
  13.1 Significance Check 243
  13.2 Discovery Potential 244

14 Summary and Outlook   247

III Appendix   251

A Breakdown of Monte-Carlo Contributions   251

B List of Monte Carlo Samples   258

C Additional Plots   277

List of Tables

1	The Gell-Mann matrices	21
2	Corrections to the top quark mass	35
3	Branching ratios of the decay modes of top quark pair production	46
4	EM shower variables used for the electron identification	98
5	Photon shower variables used for photon identification	109
6	Description of the data periods	123
7	$t\bar{t}\gamma$ event selection and selection efficiencies	140
8	Event yields before and after b-tagging	161
9	Event yields of the various $t\bar{t}$ samples before and after b-tagging	161
10	Breakdown of systematic uncertainties of the $e \to \gamma$ fake rate scale factors	195
11	Event yields of the W+jets+γ background estimation	198
12	Fraction of prompt photons in the multijet control region	204
13	Multijet background yields in the SR	204
14	Contributions of remaining MC@NLO $t\bar{t}$ events after the $t\bar{t}\gamma$ signal phase space overlap removal	208
15	Mis-identified electrons in MC@NLO $t\bar{t}$ events	209
16	Z+jets+γ, di-boson and single top background contributions	210
17	Relevant processes included in the template fit	215
18	Uncertainties of $t\bar{t}$ MC modeling	234
19	Breakdown of the final systematic uncertainties	242
20	Composition of the W+jets contributions (e+jets channel)	252
21	Composition of the W+jets contributions (μ+jets channel)	253
22	Composition of the Z+jets contributions (e+jets channel)	254
23	Composition of the Z+jets contributions (μ+jets channel)	255

24	Composition of the di-boson contributions (e+jets channel)	256
25	Composition of the di-boson contributions (μ+jets channel)	256
26	Composition of the single top contributions (e+jets channel)	257
27	Composition of the single top contributions (μ+jets channel)	257
28	$t\bar{t}\gamma$ signal sample.	259
29	Baseline $t\bar{t}$ sample.	259
30	AcerMC $t\bar{t}$ sample.	260
31	POWHEG $t\bar{t}$ samples.	260
32	JF17 di-jet sample.	260
33	AcerMC $t\bar{t}$ samples with ISR/FSR variations.	261
34	$W \to e\nu$ + jets (0...5 additional partons).	262
35	$W \to \mu\nu$ + jets (0...5 additional partons).	263
36	$W \to \tau\nu$ + jets (0...5 additional partons).	264
37	W + jets + γ samples (0...5 additional partons).	265
38	$Z \to e^+e^-$ + jets (0...5 additional partons).	266
39	$Z \to \mu^+\mu^-$ + jets (0...5 additional partons).	267
40	$Z \to \tau^+\tau^-$ + jets (0...5 additional partons).	268
41	$W + b\bar{b}$ + jets (0...3 additional partons).	269
42	$W + c\bar{c}$ + jets (0...3 additional partons).	270
43	$W + c(\bar{c})$ + jets (0...4 additional partons).	271
44	$Z \to e^+e^- + b\bar{b}$ + jets (0...3 additional partons).	272
45	$Z \to \mu^+\mu^- + b\bar{b}$ + jets (0...3 additional partons).	273
46	$Z \to \tau^+\tau^- + b\bar{b}$ + jets (0...3 additional partons).	274
47	Di-boson ($WW/WZ/ZZ$) samples.	275
48	Single top quark samples.	276

List of Figures

1	Feynman diagram of the Fermi four-point interaction model	8
2	Illustration of the SSB mechanism	15
3	Fermions and gauge bosons of the Standard Model	26
4	Unitarity triangle	28
5	History of limit settings and direct measurements of the top quark	33
6	One-loop corrections contributing to M_Z and M_W	36
7	Illustration of the MSTW2008 PDF	40
8	Feynman diagrams for single top quark production	42
9	Feynman diagrams for top quark pair production	43
10	Experimental results for the cross section of top quark pair production at Tevatron and the LHC	46
11	Representative Feynman diagram of the semi-leptonic $t\bar{t}$ decay	52
12	Representative Feynman diagrams of radiative top quark production	53
13	Representative Feynman diagrams of radiative top quark decay	54
14	View into the LHC tunnel	56
15	Schematic overview of the LHC	57
16	LHC dipole	58
17	The ATLAS detector	67
18	Particle detection	68
19	ATLAS coordinate system	69
20	The inner detector	70
21	The pixel detector	71
22	Schematic overview of the ATLAS calorimetry system	74
23	Accordion structure of the LAr calorimeter	75
24	Granularity of the three layers of the LAr EMC	76

LIST OF FIGURES

25	View into the ATLAS detector	81
26	The ATLAS end-cap toroid	82
27	Schematic overview of the ATLAS trigger and DAQ system	83
28	Identification efficiencies of unconverted and converted photons ($|\eta| < 0.6$)	110
29	Identification efficiencies of unconverted and converted photons ($0.6 \leq |\eta| < 1.37$)	111
30	Identification efficiencies of unconverted and converted photons ($1.52 \leq |\eta| < 1.81$)	112
31	Identification efficiencies of unconverted and converted photons ($1.81 \leq |\eta| < 2.37$)	113
32	Integrated and instantaneous luminosity in 2011	121
33	Integrated and instantaneous luminosity in 2011	122
34	Discrepancies of photon shower shape variables between data and MC simulation	131
35	Ratio of photon transverse energy and jet transverse momentum as a function of ΔR	136
36	Flowchart of the ATLAS event simulation chain	143
37	Transverse momenta of WHIZARD truth photons	145
38	Illustration of the MC@NLO $t\bar{t}$ phase space leakage (1)	152
39	Illustration of the MC@NLO $t\bar{t}$ phase space leakage (2)	153
40	Average number of bunch crossings in data and MC simulation	155
41	Comparison between data and MC simulation before b-tagging (1)	162
42	Comparison between data and MC simulation before b-tagging (2)	163
43	Comparison between data and MC simulation before b-tagging (3)	164

44	Comparison between data and MC simulation before b-tagging (4)	165
45	Comparison between data and MC simulation after b-tagging (1)	166
46	Comparison between data and MC simulation after b-tagging (2)	167
47	Comparison between data and MC simulation after b-tagging (3)	168
48	Comparison between data and MC simulation after b-tagging (4)	169
49	Sources of hadron fakes	171
50	Photon isolation shapes for signal and background MC simulation	174
51	Photon selection efficiency vs. background rejection for various isolation definitions	175
52	Invariant two-electron mass	178
53	Comparison between electron and photon isolation	179
54	Electron track isolation in data for different bins in p_T and $|\eta|$	180
55	Isolation spectrum of hadron fakes	182
56	Distribution of transverse momenta and pseudo-rapidities of hadron fakes	184
57	Final prompt photon and hadron fake isolation templates	185
58	Electron efficiencies obtained from MC simulation	190
59	Invariant mass distributions of ee and $e\gamma_{\text{fake}}$ events	193
60	Comparison of the $e \to \gamma$ fake rate in 2×3 bins in photon p_T and $|\eta_{S2}|$	194
61	$e \to \gamma$ fake rate measured in three bins of number of primary vertices	196
62	Scale factors derived from the $e \to \gamma$ fake rate measurement	197

LIST OF FIGURES xiii

63 Fraction of prompt photons f_γ obtained from template fits in the CR for events containing one, two and three jets 200

64 Template fit of W+jets+γ in the CR 201

65 Template fits in the CR for obtaining $f_\gamma^{\mathrm{prompt}}$ 205

66 Comparison of the prompt photon and hadron fake templates between data and MC simulation (1) 211

67 Comparison of the prompt photon and hadron fake templates between data and MC simulation (2) 212

68 Comparison of the prompt photon and hadron fake templates between data and MC simulation (3) 213

69 Result of the template fit 216

70 Marginalized distribution of expected number of $t\bar{t}\gamma$ candidates 217

71 Pseudo data 221

72 Distribution of the expected number of signal events for 3000 pseudo experiments (1) 222

73 Distribution of the expected number of signal events for 3000 pseudo experiments (2) 223

74 Summary plots of JES uncertainties 226

75 Illustration of the determination of the overall PDF uncertainty 233

76 Systematic variations of the prompt photon and hadron fake template 239

77 Event selection efficiency of the $t\bar{t}\gamma$ signal sample evaluated for different numbers of average bunch crossings 241

78 Measured and expected significance 245

79 Discovery potential of the process $t\bar{t}\gamma$ 246

80 Comparison between data and MC simulation before b-tagging (1) 278

81	Comparison between data and MC simulation before b-tagging (2)	279
82	Comparison between data and MC simulation before b-tagging (3)	280
83	Comparison between data and MC simulation before b-tagging (4)	281
84	Comparison between data and MC simulation before b-tagging (5)	282
85	Comparison between data and MC simulation before b-tagging (6)	283
86	Comparison between data and MC simulation after b-tagging (1)	284
87	Comparison between data and MC simulation after b-tagging (2)	285
88	Comparison between data and MC simulation after b-tagging (3)	286
89	Comparison between data and MC simulation after b-tagging (4)	287
90	Comparison between data and MC simulation after b-tagging (5)	288
91	Comparison between data and MC simulation after b-tagging (6)	289
92	Comparison between data and MC simulation after b-tagging (7)	290
93	Comparison between data and MC simulation after b-tagging (8)	291
94	Invariant mass spectra $Z \to e^+e^-$ and $Z \to e\gamma_{\text{fake}}$ in bins of p_T and $|\eta|$ (1)	292
95	Invariant mass spectra $Z \to e^+e^-$ and $Z \to e\gamma_{\text{fake}}$ in bins of p_T and $|\eta|$ (2)	293

96	Invariant mass spectra $Z \to e^+e^-$ and $Z \to e\gamma_{\text{fake}}$ in bins of p_T and $|\eta|$ (3)	294
97	Template fit of $W + \text{jets} + \gamma$ in the CR for different jet multiplicities (1)	295
98	Template fit of $W + \text{jets} + \gamma$ in the CR for different jet multiplicities (2)	296
99	Template fit of $W + \text{jets} + \gamma$ in the CR for different jet multiplicities (3)	297
100	Systematic uncertainties related to jets (1)	298
101	Systematic uncertainties related to jets (2)	299
102	Systematic uncertainties related to b-tagging performance	299
103	Systematic uncertainties related to muon performance (1)	300
104	Systematic uncertainties related to muon performance (2)	301
105	Systematic uncertainties related to electron performance (1)	302
106	Systematic uncertainties related to electron performance (2)	303
107	Systematic uncertainties related to $\slashed{E}_T$	304
108	Systematic uncertainties related to the EMC hardware failure	304
109	Systematic uncertainties related to photon performance	305
110	Systematic uncertainties related to the template modeling	306
111	Systematic uncertainties related to MC simulation (1)	307
112	Systematic uncertainties related to MC simulation (2)	308
113	Systematic uncertainties related to background estimation (1)	309
114	Systematic uncertainties related to background estimation (2)	310

List of Acronyms

AOD	Analysis Object Data
ATLAS	A Torodial Large ApparatuS
a.u.	arbitrary units
BR	Branching Ratio
CAD	Computer Aided Design
CERN	Conseil Europeén de la Recherche Nucléaire[1]
CKM	Cabbibo Kobaiashi Maskawa (matrix)
CMS	Center of Mass System or
	Compact Muon Solenoid
CP	Charge-Parity (invariance)
CR	ControlRegion
CS	CryoStat
CSC	Cathode Strip Chamber
DAQ	Data AcQisition
DPD	Derived Physics Data
DY	Drell-Yan (process)
EF	Event Filter (trigger)
EMC	Electro-Magnetic Calorimeter
EMEC	Electro-Magnetic End-Cap calorimeter
ESD	Event Summary Data
EW	Electro-Weak
EWSB	Electro-Weak Symmetry Breaking
FCAL	Forward CALorimeter
FCNC	Flavor Changing Neutral Current
HF	Heavy Flavor
IBS	Intra-Beam Scattering
ID	Inner Detector or
	IDentification (electron/photon)

LIST OF ACRONYMS

IP	Interaction Point
ITC	Intermediate Tile Calorimeter
KS	Kolmogorov-Smirnov (test)
L1	Level 1 (trigger)
L2	Level 2 (trigger)
LAr	Liquid Argon
LB	Lumi Block
LCG	LHC Computing Grid
HC	Hadronic Calorimeter
HEC	Hadronic End-Cap calorimeter
LEP	Large Eadron Positron Collider
LHC	Large Hadron Collider
LO	Leading Order (calculation)
MDT	Monitored Drift Tube
MVA	Multi-Variate Analysis
NLO	Next-to-Leading Order
NNLO	Next-to-Next-to-Leading Order
PDF	Parton Distribution Function
PMT	Photo Multiplier Tube
PS	Parton Showering
RF	Radio Frequency
RMS	Root Main Square
ROB	Read Out Buffer
ROD	Read Out Driver
RoI	Region of Interest
RPC	Resistive Plate Chamber
SCT	Semi-Conductor Tracker
SF	Scale Factor
SM	Standard Model
SR	Signal Region
SSB	Spontaneous Symmetry Breaking

TGC	Thin Gap Chamber
TRT	Transition Radiation Tracker
UE	Underlying Event
vdM	van der Meer
VEV	Vacuum Expectation Value
WLS	WaveLength Shifting

---

[1]Meanwhile, the official denotation is "European Organization for Nuclear Research" or, in French, "Organisation Européenne de la Recherche Nucléaire". The acronym "CERN" was kept for historical reasons.

0 Preface

With the discovery of the top quark at the Tevatron collider in 1995, the existence of the last of the six quarks of the Standard Model of Particle Physics could be verified experimentally.

Since that time, the Tevatron significantly increased the amount of recorded data and thus many important properties of the top quark could be measured at a high precision. The mass of the top quark has been determined at an overall uncertainty smaller than 1 % and also the measured cross sections of several production and decay channels could be tested successfully against the theoretical predictions.

Nevertheless, there are still a lot of unanswered questions; many important parameters and properties of the top quark could not be determined at high precision or even not at all.

Since the top quark is the only quark in the Standard Model that decays before it can hadronize, it provides a "single quark laboratory", excelling it compared to the five other quarks. This unique feature makes it possible to study the coupling structure of gauge bosons to the top quark and to quarks in general and allows for the long-term measurement of the vertex structures tg, tZ and $t\gamma$.

Especially the couplings to the electroweak gauge bosons Z and γ could reveal a possible inner structure of quarks, being a consequence of several New Physics models, which would emerge as additional electromagnetic or weak dipole moments. Precision measurements at LEP could only weakly constrain limits on such exotic dipole moments.

The LHC collider at CERN, near Geneva, is a "top quark factory" which allows to study and answer such open question in the near future. The analysis of the $t\gamma$ vertex structure is a first step into this direction since this process provides the largest cross section among top quark gauge boson couplings and has a relatively unique event signature.

The measurement of the absolute cross section of the inclusive semileptonic top quark pair decay mode with the simultaneous emission of a photon has been performed by the ATLAS experiment and is the first LHC measurement of that process.

Part I
Introduction

1 The Standard Model of Particle Physics

The Standard Model of Particle Physics is a successful theoretical description of the elementary particles and the interactions among them. It has been developed over the past 50 years and it provides a clear gauge structure at highly predictive power.

1.1 Gauge Groups and Symmetries

In modern high energy particle physics, elementary particle and all interactions among them are highly relativistic. Hence it is necessary to describe a theory within the framework of the Lorentz group or, more general, the Poincaré group.
To be able to embed the components of the Standard Model into the symmetry of a group, there are two constraints on the theories [1, p. 34]:

1. All fields have to transform as irreducible representations of the Lorentz and the Poincaré group plus some isospin group

2. The theory has to be unitary and the actions have to be causal, renormalizable and invariant under transformations under these groups.

1.2 Quantum Field Theories (QFT)

In the beginning of the 20th, two markable changes in the scientific view onto the structure and behavior of matter appeared. The one change was the extension of the Newtonian mechanics to relativistic mechanics by A. EINSTEIN in 1905, merging energy and matter via his famous formula $E = mc^2$ to an equivalent by linking them just by the constant speed of light.

The second milestone was the invention of quantum mechanics, first developed as a transition from classical mechanics to a wave equation by the Schrödinger equation:

$$i\hbar\frac{\partial}{\partial t}\psi(\vec{x},t) = \hat{H}\psi(\vec{x},t) \qquad (1)$$

describing the time evolution of a particle with the wave function $\psi(\vec{x},t)$ by the Hamilton operator $\hat{H}$.

In the 20ies of the 20th century, the problem emerged describing photons with classical quantum mechanics. PAUL DIRAC described "wave quanta" of photons for the first time in 1927 [2]. With this first description of a quantized electro-magnetic field he already could describe the relativistic contributions to the hydrogen spectra as well as the spin and magnetic moment of the electron.

But it turned out soon that with this early version of the QED, only the lowest order of perturbation could be described, the calculation of higher order contributions lead to infinities. The reason originated from calculating integrals for $x \to 0$ ($k \to \infty$) [1, p. 4]. This problem was solved 1949 by renormalization [1, p. 5]. Since then, QFTs were successfully applied to describe electro-magnetic, weak and strong interactions.

1.2.1 Lagrangian and Equation of Motion

In QFTs, physical systems are not defined by a set of generalized coordinates q^i and p^i but by fields ϕ which have an infinite number of

1 THE STANDARD MODEL OF PARTICLE PHYSICS

degrees of freedom which is known as *second quantization* [1, p. 21]. This leads to the transitions

$$x^i \to \phi(x_\mu),$$
$$q^i \to \partial_\mu \phi(x_\mu).$$

Usually, QFTs are defined by their Lagrangian density rather than by their Hamiltonian. Still, the QFT Lagrangians obey the Action Principle

$$\delta S = \int \mathrm{d}^4 x \left(\frac{\delta \mathscr{L}}{\delta \phi} \delta \phi + \frac{\delta \mathscr{L}}{\delta \partial_\mu \phi} \delta \phi \right) = 0. \qquad (2)$$

The QFT Euler-Lagrange equations of motions are hence given by

$$\partial_\mu \frac{\delta \mathscr{L}}{\delta \partial_\mu \phi} = \frac{\delta \mathscr{L}}{\delta \phi}. \qquad (3)$$

1.2.2 Neutral Scalar Field

The simplest QFT is the relativistic, neutral scalar field, described by the Lagrangian

$$\mathscr{L} = \frac{1}{2}(\partial_\mu \phi)^2 - \frac{1}{2} m^2 \phi^2 \qquad (4)$$

with the solution

$$(\Box - m^2)\psi = 0, \qquad (5)$$

called Klein-Gordon equation. For a charged scalar field, ϕ has to be split up into a linear combination of two scalar fields:

$$\phi = \frac{1}{\sqrt{2}}(\phi_1 + i\phi_2). \qquad (6)$$

The corresponding Lagrangian is

$$\mathscr{L} = \partial_\mu \phi^\dagger \partial^\mu \phi - m^2 \phi^\dagger \phi. \qquad (7)$$

The solutions for ϕ can be decomposed via superpositions of ladder operators $a_i(k)$:

$$\phi_i = \int \frac{d^3k}{\sqrt{(2\pi)^3 2\omega_k}} \left(a_i(k)e^{-ikx} + a_i^\dagger(k)e^{ikx} \right) \quad (i = 1, 2). \qquad (8)$$

If the representation

$$a(k) = \frac{1}{\sqrt{2}}[a_1(k) + ia_2(k)] \qquad b(k) = \frac{1}{\sqrt{2}}[a_1(k) - ia_2(k)] \qquad (9)$$

is chosen, one obtains

$$Q = J_0 = \int d^3x\, i(\phi^\dagger \partial_0 - \partial_0 \phi^\dagger \phi) = N_a - N_b \qquad (10)$$

for the Noether current $J_\mu = i\phi^\dagger \partial_\mu \phi - i\partial_\mu \phi^\dagger \phi$. The positive sign of N_a and the negative sign of N_b can be interpreted as electric charge [1, p. 71].

1.2.3 The Dirac Field

Since the Klein-Gordon equation (5) is quadratic in time, it can have negative solutions, leading to negative probabilities. DIRAC translated Eq. (5) into a set of linear equations using spinors, leading to the Dirac field which is capable of describing spin $\frac{1}{2}$ particles.

From comparing the linear equation

$$i\frac{\partial \psi}{\partial t} = \left(-i\alpha_i \nabla^i + \beta m\right)\psi \qquad (11)$$

with the desired, final quadratic form

$$-\frac{\partial^2}{\partial t^2}\psi = \left(-i\alpha \cdot \nabla + \beta m\right)^2 \psi, \qquad (12)$$

1 THE STANDARD MODEL OF PARTICLE PHYSICS

it can be deduced that this set of equations can only be solved if the matrices α_i and β satisfy

$$\{\alpha_i, \alpha_k\} = 2\delta_{ik}, \quad (13)$$
$$\{\alpha_i, \beta\} = 0, \quad (14)$$
$$\alpha_i^2 = \beta^2 = 1. \quad (15)$$

By defining the *gamma matrices* $\gamma^{0...4}$ by

$$\beta = \begin{pmatrix} 1 & 0 \\ 0 & -1 \end{pmatrix} \quad \alpha^i = \begin{pmatrix} 0 & \sigma^i \\ \sigma^i & 0 \end{pmatrix} \quad (16)$$

$$\Rightarrow \quad \gamma^0 = \begin{pmatrix} 1 & 0 \\ 0 & -1 \end{pmatrix} \quad \gamma^i = \begin{pmatrix} 0 & \sigma^i \\ -\sigma^i & 0 \end{pmatrix} \quad (17)$$

with the Pauli matrices $\sigma^{1...3}$, (11) can be written as the *Dirac equation*:

$$(i\gamma^\mu \partial_\mu - m)\psi = 0. \quad (18)$$

The gamma matrices fulfil the anti-commutator relation

$$\{\gamma^\mu, \gamma^\nu\} = 2g^{\mu\nu}. \quad (19)$$

The hermitian conjugate of (18) is given by

$$\bar{\psi}(i\gamma^\mu \overleftarrow{\partial}_\mu + m) = 0 \quad (20)$$

with the hermitian conjugate of the Dirac field $\bar{\psi} = \psi^\dagger \gamma^0$. The Lagrangian of the free Dirac field is given by

$$\mathscr{L} = \bar{\psi}(i\slashed{\partial} - m)\psi. \quad (21)$$

1.3 The Weak Force

In 1914, J. CHADWICK observed that the electron momentum spectrum of the β decay was continuous. This only could be explained by the existence of a second, invisible particle that could take away some momentum in order to conserve momentum and energy conservation of the β decay [3].

In 1930, W. PAULI proposed the existence of a weakly interacting particle in an open letter to explain the continuous β spectrum [4] which he in 1934 called "neutrino" [5].

FERMI described the β decay as the product of two independent neutral currents [6] interacting at a four-point vertex (Fig. 1)

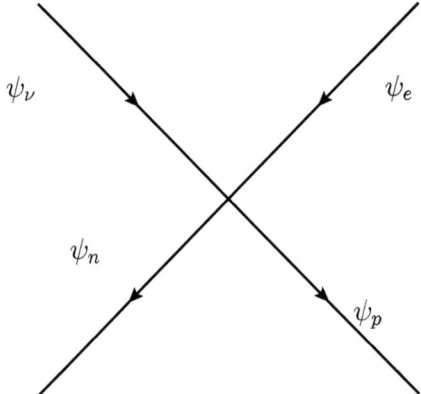

Figure 1: Feynman diagram of the Fermi four-point interaction model.

$$\mathscr{L}_\beta = \frac{G_F}{\sqrt{2}}(\bar{\psi}_p \gamma^\mu \psi_n)(\bar{\psi}_e \gamma_\mu \psi_\nu) \tag{22}$$

with the *weak currents* $J_W^\mu = \bar{\psi}\gamma^\mu\psi$. GAMOV and TELLER proposed an extension of that structure to describe also transitions with $\Delta J_{\text{nucl.}} \neq 0$ [7]:

$$\mathscr{L}_{\text{weak}} = \frac{G_F}{\sqrt{2}} \sum_i C_i (\bar{\psi}_p \Gamma^i \psi_n)(\bar{\psi}_e \Gamma^i \psi_\nu) , \tag{23}$$

1 THE STANDARD MODEL OF PARTICLE PHYSICS

where Γ^μ takes into account vectorial and axial-vectorial couplings that consider any kind of nuclear transitions ($\Delta J = 0, \pm 1$ and $\Delta J = 0, 0 \to 0$):

$$\Gamma^\mu_S = 1\,, \quad \Gamma^\mu_P = \gamma_5\,, \quad \Gamma^\mu_V = \gamma^\mu\,, \quad \Gamma^\mu_A = \gamma^\mu\gamma_5\,, \quad \Gamma^{\mu\nu}_T = \sigma^{\mu\nu}\,. \quad (24)$$

The Fermi model of nuclear transitions was extended in 1957 by the coupling $\bar\psi\Gamma(1\pm\gamma_5)\psi$ after parity violation had been discovered in β decays by WU et al. in the same year [8].

In the same year, SCHWINGER, LEE and YANG proposed the description of the weak force by the exchange of a (massive) vector boson [9, 10] in order to restore unitarity for high interaction energies since the Fermi theory would spoil unitarity conservation for CMS energies of $\sqrt{s} \gtrsim 300\,\text{GeV}$. In this theory, a vector boson connects the nuclear currents by a propagator that avoids divergent cross sections:

$$\mathscr{L}_F = \tfrac{G_F}{\sqrt{2}} J^\mu(p) J_\mu(e)$$
$$\to [iG_W J^\mu(p)] \left[\tfrac{i}{M_W^2 - k^2}\left(g_{\mu\nu} - \tfrac{k_\mu k_\nu}{M_W^2}\right)\right] [iG_W J^\nu(e)] \quad (25)$$

with the weak coupling constant G_W which can be calculated from the Fermi constant G_F by comparing Eq. (23) and Eq. (25) in the low energy limit:

$$G_W^2 = \frac{M_W^2 G_F}{\sqrt{2}}\,. \quad (26)$$

1.4 Local Gauge Invariance and Quantum Electrodynamics

Symmetries play an important role in physics. If a symmetry can be found for a physical system or within a theory for that system, the behavior under symmetry transformations directly leads to a quantity that is conserved under that transformation (*Noether theorem*).

As a consequence, a Lagrangian can be modified in such a way that the action remains the same for the original and the modified Lagrangian. This modification is called *gauge*.

There are two general classes of gauge symmetries: *Global gauge invariance* and *local gauge invariance*. The difference between both is that global gauge invariance is achieved by global transformations and local gauge invariance by transformations that depend on the space-time point of the considered physical system.

Since a real physical system, especially in high energy environments, has to obey the theory of special relativity, transformations will look different in different space-time points but have to yield the same physical results. Therefore, any meaningful theory of particle physics has to be locally gauge invariant.

If an action is invariant under some group of transformations (symmetry), then there exist one or more conserved quantities (constants of motion) which are associated to these transformations [11].

This implies that the Lagrangian of the free Dirac field (Eq. (21)) has no conserved quantity since it is not invariant under local space-time transformations. Taking the series of infinitesimal transformations $\exp(-i\alpha(x))$, i. e. $\psi' = \exp(-i\alpha(x))\psi$, $\mathscr{L}$ transforms as

$$\mathscr{L}' = \mathscr{L} + \bar{\psi}\gamma_\mu\psi\partial^\mu\alpha(x) \,. \tag{27}$$

Local gauge invariance can be restored by applying the covariant derivative instead of ∂:

$$D_\mu = \partial_\mu + ieA_\mu \tag{28}$$

and requiring that $A'_\mu = A_\mu + \frac{1}{e}\partial_\mu\alpha(x)$. Restoring local gauge invariance introduces a new term in Eq. (21) which arises from the additional term

$$\mathscr{L}' = \mathscr{L} - e\bar{\psi}\gamma_\mu\psi A^\mu \tag{29}$$

1 THE STANDARD MODEL OF PARTICLE PHYSICS

in the covariant derivative and hence introduces the coupling between external Dirac fields ψ (e. g. electrons) and the photon field A_μ. By adding the quantized electro-magnetic field as interaction, the Theory of Quantum Electrodynamics (QED) is obtained. Its Lagrange density is

$$\mathscr{L}_{\text{QED}} = -\frac{1}{4}F_{\mu\nu}F^{\mu\nu} + \sum_n \bar{\psi}_n(i\gamma^\mu D_\mu - m_n)\psi_n\,, \qquad (30)$$

where $F^{\mu\nu}$ is the field tensor of the electro-magnetic field:

$$F_{\mu\nu} = \partial_\mu A_\nu - \partial_\nu A_\mu\,. \qquad (31)$$

It should be noted that the mass term in Eq. (18) has to be set to $m = 0$. Otherwise, terms of the form $m\bar{\psi}\psi$ would be inserted in the Lagrangian that would destroy the principle of local gauge invariance. Generally, gauge fields have to be massless in local gauge invariant theories.

1.5 The GSW Theory

In the beginning of the 1960ies, when the description of the QED with Dirac fields already had successfully predicted a lot of experimental results, GLASHOW, SALAM and WEINBERG were trying to combine electro-magnetic and weak interactions in one theory (GSW theory). The basic ansatz was to assume that both the weak and EM force should be the result of a *spontaneous symmetry breaking*, resulting into some massive vector bosons and a massless photon.

Using a $SU(2)$ theory, particles can be combined in left-handed doublets an right-handed singlets [1, p. 335]:

$$L = \begin{pmatrix} \nu_e \\ e \end{pmatrix}_L \qquad R = (e)_R\,. \qquad (32)$$

A $SU(2)$ theory obeys the Lie Algebra with the commutation relation

$$[\tau_a, \tau_b] = f_{abc}\tau_c, \qquad (33)$$

where $\tau_{a,b,c}$ are the Pauli matrices and the *generators* of the $SU(2)$.

The $SU(2)$ gauge fields are named $W_{\mu\nu}^a$ ($a = 1\ldots 3$) and those of $U(1)$ as $B_{\mu\nu}$. The gauge fields are derived from the potentials via

$$B_{\mu\nu} = \partial_\mu B_\nu - \partial_\nu B_\mu, \qquad (34)$$
$$W_{\mu\nu}^a = \partial_\mu W_\nu^a - \partial_\nu W_\mu^a + g f_{abc} W_\mu^b W_\nu^c. \qquad (35)$$

With this convention, the gauge field self-interaction part of the Lagrangian $\mathscr{L}_{\text{gauge}}$ can be written as [12] [1, p. 336]

$$\mathscr{L}_{\text{gauge}} = -\frac{1}{4} W_{\mu\nu}^a W^{\mu\nu,a} - \frac{1}{4} B_{\mu\nu} B^{\mu\nu} \qquad (36)$$

and the complete Lagrangian reads

$$\mathscr{L}_{SU(2)\times U(1)} = \mathscr{L}_{\text{gauge}} + \mathscr{L}_f + \mathscr{L}_\phi + \mathscr{L}_{\text{Yuk}}, \qquad (37)$$

where $\mathscr{L}_f$ the kinetic energy of the free Dirac particles, $\mathscr{L}_\phi$ the Lagrangian of the Higgs field self-interaction (see Sec. 1.6) and $\mathscr{L}_{\text{Yuk}}$ the Lagrangian of fermions interacting with the Higgs field (*Yukawa coupling*, see Sec. 1.6.3).

$B_{\mu\nu}$ and $W_{\mu\nu}^a$ are not the field strength tensors of the physical gauge field; only after electroweak symmetry breaking (EWSB) they recom-

1 THE STANDARD MODEL OF PARTICLE PHYSICS

bine as the physical photon field A_μ, the two charged W bosons and the electrically neutral Z boson [1, p. 337]:

$$Z_\mu = \frac{gW_\mu^3 + g'B_\mu}{\sqrt{g^2 + g'^2}} =: \cos\theta_w W_\mu^3 + \sin\theta_w B_\mu, \tag{38}$$

$$A_\mu = \frac{-g'W_\mu^3 + gB_\mu}{\sqrt{g^2 + g'^2}} =: \sin\theta_w W^3\mu + \cos\theta_w B_\mu, \tag{39}$$

$$W_\mu^\pm = \frac{1}{\sqrt{2}}(W_\mu^1 \pm iW_\mu^2), \tag{40}$$

with the *electroweak mixing angle* or *Weinberg angle* θ_w which is defined as

$$\cos\theta_w = \frac{g}{\sqrt{g^2 + g'^2}}, \tag{41}$$

$$\tan\theta_w = \frac{g'}{g}. \tag{42}$$

1.6 The Higgs Mechanism

In a locally gauge invariant theory, no explicit masses of fields are allowed since they would spoil local gauge invariance. Gauge boson masses are introduced by spontaneous symmetry breaking (SSB). The basic idea is that the lowest energy (vacuum) state does not respect the gauge symmetry and induces effective masses for particles propagating through it [12].

1.6.1 Spontaneous Symmetry Breaking

Given a complex scalar field ϕ, the vacuum expectation value (VEV) is given by

$$v = \langle 0|\phi|0\rangle = \text{const.} \tag{43}$$

ϕ is symmetric under the discrete transformation $\phi \to \phi' = -\phi$. If the potential of ϕ is labeled as $V(\phi)$, the Lagrangian can be written as

$$\mathscr{L}_\phi = \frac{1}{2}\partial_\mu\phi\partial^\mu\phi - V(\phi). \qquad (44)$$

An arbitrary $V(\phi)$ can be expanded in orders of ϕ:

$$V(\phi) = \frac{1}{2}\mu^2\phi^2 + \frac{1}{4}\lambda\phi^4. \qquad (45)$$

(44) is invariant under the transformation $\phi \to -\phi$ whereas the VEV is not necessarily: Considering the Hamiltonian $\mathscr{H}_\phi = \frac{1}{2}[(\partial_0\phi)^2 + (\nabla\phi)^2] + V(\phi)$, the vacuum state ϕ_0 of ϕ can be written as

$$\phi_0(\mu^2 + \lambda\phi_0^2) = 0. \qquad (46)$$

$\mathscr{H}_\phi$ should be bounded, so $\lambda > 0$ is required; thus the minimum of $V(\phi)$ depends on the sign of μ^2: for $\mu^2 > 0$, there is only one vacuum state (at $\phi = 0$), for $\mu^2 < 0$, two states emerge at $\phi^\pm = \sqrt{-\mu^2/\lambda}$ (see Fig. 2). This means that, although $\mathscr{L}_\phi$ is invariant whether choosing $\phi = \phi^+$ or $\phi = \phi^-$, but the vacuum state is not: the symmetry is spontaneously broken.

Eq. (44) can be written in terms of a new potential $\tilde{\phi}$ which represents the location of the new vacuum state after symmetry breaking:

$$\tilde{\phi} = \phi - v. \qquad (47)$$

With this choice, the Lagrangian becomes

$$\mathscr{L}_{\tilde{\phi}} = \frac{1}{2}\partial_\mu\tilde{\phi}\partial^\mu\tilde{\phi} - \frac{1}{2}\sqrt{-2\mu^2}\tilde{\phi}^2 - \lambda v\tilde{\phi}^3 - \frac{1}{4}\lambda\tilde{\phi}^4. \qquad (48)$$

$\tilde{\phi}$ has obtained the mass $M_{\tilde{\phi}} = \sqrt{-2\mu^2}$. The original symmetry has vanished due to the $\tilde{\phi}^3$ term, i.e. there is no $\tilde{\phi} \to -\tilde{\phi}$ symmetry anymore.

1 THE STANDARD MODEL OF PARTICLE PHYSICS 15

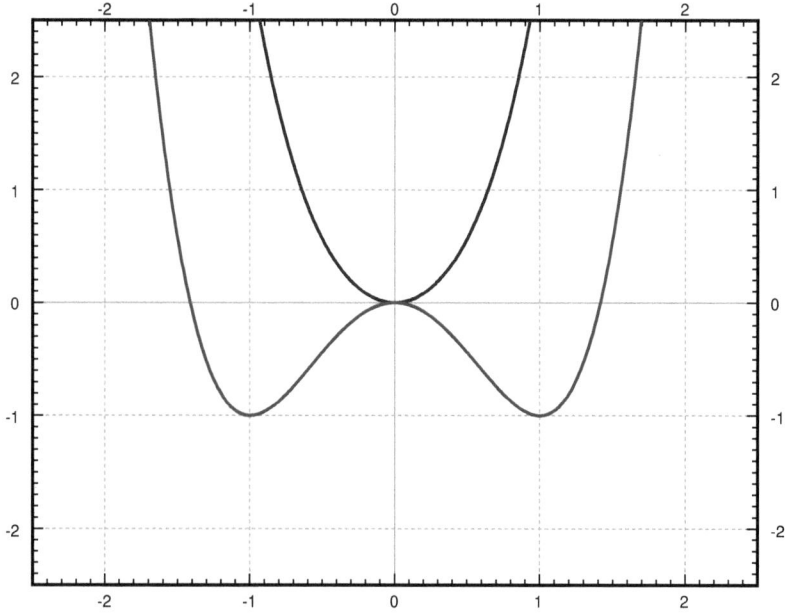

Figure 2: Illustration of the SSB mechanism, by means of $V(\phi)$, shown for $\mu^2 > 0$ (blue) and $\mu^2 < 0$ (green). In this example, $\lambda = 0.25$ and $\mu^2 = \pm 1$ have been chosen.

If a continuous symmetry is broken, the Lagrangian with the shifted fields will contain one massless boson for each broken generator, the *Nambu-Goldstone bosons*.

1.6.2 The Higgs Mechanism in the Standard Model

In order to establish spontaneous symmetry breaking in the GSW theory, a doublet Φ of two scalar fields ϕ^+ and ϕ^0 is defined [11][pp. 44]:

$$\Phi = \begin{pmatrix} \phi^+ \\ \phi^0 \end{pmatrix}. \qquad (49)$$

The corresponding Lagrangian is

$$\mathscr{L}_{\text{Higgs}} = \frac{1}{2}\partial_\mu \Phi^\dagger \partial^\mu \Phi - V(\Phi^\dagger \Phi),\qquad(50)$$

with the Higgs potential $V(\Phi^\dagger\Phi)$:

$$V(\Phi^\dagger\Phi) = \mu^2 \Phi^\dagger\Phi + \lambda(\Phi^\dagger\Phi)^2.\qquad(51)$$

The VEV of the broken symmetry is usually defined as

$$\langle\Phi\rangle_0 = \begin{pmatrix} 0 \\ v/\sqrt{2} \end{pmatrix}\qquad(52)$$

with $v = \sqrt{-\frac{\mu^2}{\lambda}}$. The $SU(2)_L \times U(1)_Y$ GSW theory has to contain the exact QED after symmetry breaking, therefore the global symmetry $\langle\Phi\rangle_0 \to \langle\Phi\rangle'_0 = \exp(i\alpha Q)\langle\Phi\rangle_0$ has to be conserved. With the approximation $\exp(i\alpha Q) \simeq 1 + i\alpha Q$ and the Gell-Mann-Nishima formula $Q = T_3 - \frac{1}{2}Y$, applying the operator Q on $\langle\Phi\rangle_0$ must not have an effect, which is indeed the case:

$$\begin{aligned} Q\langle\Phi\rangle_0 &= \left(T_3 + \frac{1}{2}Y\right)\langle\Phi\rangle_0 \\ &= \frac{1}{2}\left[\begin{pmatrix} 1 & 0 \\ 0 & -1 \end{pmatrix} + \begin{pmatrix} 1 & 0 \\ 0 & 1 \end{pmatrix}\right]\begin{pmatrix} 0 \\ v/\sqrt{2} \end{pmatrix} = 0. \end{aligned}$$

The other gauge bosons that originate from the remaining broken generators T_1, T_2 and $T_3 - Y/2$ should obtain a mass from the SSB. For this purpose, Φ is expanded from the vacuum state by an excitation field H:

$$\Phi = \exp\left(i\frac{\tau^i \chi^i}{2\,v}\right)\begin{pmatrix} 0 \\ (v+H)/\sqrt{2} \end{pmatrix} \simeq \frac{1}{\sqrt{2}}\begin{pmatrix} i\sqrt{2}\omega^+ \\ v + H - iz^0 \end{pmatrix}\qquad(53)$$

1 THE STANDARD MODEL OF PARTICLE PHYSICS

with ω^+ and z^0 being the Goldstone bosons. The fields χ^i can be inserted into the gauge parameter α, choosing the *unitary gauge* $\alpha^i = \chi^i/v$ eliminates the Goldstone bosons. Together with the covariant derivative of the $SU(2)_L \times U(1)_Y$

$$D_\mu = \partial_\mu + ig\frac{\tau^i}{2}W^i_\mu + i\frac{g'}{2}YB_\mu, \qquad (54)$$

Eq. (50) can be rewritten as

$$\mathscr{L}_{\text{Higgs}} = \left| \left(\partial_\mu + ig\frac{\tau^i}{2}W^i_\mu + i\frac{g'}{2}YB_\mu \right) \begin{pmatrix} 0 \\ (v+H)/\sqrt{2} \end{pmatrix} \right|^2 \qquad (55)$$
$$-\mu^2\frac{(v+H)^2}{2} - \lambda\frac{(v+H)^4}{4}.$$

By expressing B_μ and W^3_μ in terms of the physical gauge fields A_μ and Z_μ (Eq. (38)) one obtains

$$\mathscr{L}_{\text{Higgs}} = \frac{1}{2}\partial_\mu H \partial^\mu H + \frac{g^2}{4}(v+H)^2 \left(W^+_\mu W^{-\mu} + \frac{1}{2\cos\theta_w}Z_\mu Z^\mu \right). \qquad (56)$$

In this form, the quadratic mass terms of the massive gauge bosons read

$$\frac{g^2 v^2}{4} W^+_\mu W^{-\mu} \qquad \frac{g^2 v^2}{8\cos^2\theta_w} Z_\mu Z^\mu, \qquad (57)$$

so finally

$$M_W = \frac{gv}{2} \qquad M_Z = \frac{gv}{2\sqrt{2}\cos\theta_w}. \qquad (58)$$

In the low energy approximation (Fermi theory), the VEV of the Higgs field can be written explicitly:

$$v = \sqrt{\sqrt{2}G_F} \simeq 246\,\text{GeV}. \qquad (59)$$

The terms pure in H are

$$\mathscr{L}_H = -\frac{1}{2}(-2\mu^2)H^2 + \frac{1}{4}\mu^2 v^2 \left(\frac{4}{v^3}H^3 + \frac{1}{v^4}H^4 - 1\right) \quad (60)$$

and determine the Higgs boson mass to be $M_H = \sqrt{-2\mu^2}$. Furthermore, it predicts cubic and quartic Higgs interactions.

The Higgs boson could not be uniquely identified experimentally so far. Recent searches performed by ATLAS and CMS have identified a new boson X with a mass of $M_X \approx 126\,\text{GeV}$ that meets or at least does not contradict all requirements demanded for the properties the Higgs boson must have [13, 14].

1.6.3 Yukawa Couplings

The principle of spontaneous symmetry breaking a priori only explains how the massive gauge bosons achieve their mass and how these masses are related to each other. In order to give all fermions their masses, they have to be coupled to the Higgs field artificially by the Yukawa coupling. In this model, the massive fermions couple to the Higgs field ϕ like

$$\mathscr{L}_{\text{Yuk.}}^q = -Y_{ij}^d \overline{Q_{Li}^I} \phi d_{Ri}^I - Y_{ij}^u \overline{Q_{Li}^I} \epsilon \phi^* u_{Ri}^I + \text{h.c.} \quad (61)$$

where $Y^{u,d}$ are complex 3×3 matrices, ϵ is the anti-symmetric 2×2 tensor, Q_L^I are the left-handed quark doublets and u_R^I (d_R^I) are the right-handed up- (down-)type quark singlets [15]. The quark states are given in the flavor (weak) eigenstates. The mass terms for the quarks are obtained by diagonalizing $Y^{u,d}$:

$$M_{\text{diag}}^{u,d} = V_L^{u,d} Y^{u,d} V_R^{\dagger(u,d)} (v/\sqrt{2}) \,. \quad (62)$$

1 THE STANDARD MODEL OF PARTICLE PHYSICS

$V^{u,d}$ are given by the CKM matrix (see Sec. 1.10). Similarly, leptons are given their masses via [11]

$$\begin{aligned}\mathscr{L}^\ell_{\text{Yuk.}} &= -G_\ell \left[\bar{\ell}_R \phi \ell_L + \bar{\ell}_L \phi^* \ell_R\right] & (63)\\ &= -G_L \frac{v+H}{\sqrt{2}} \left[\bar{\ell}_R(0,1)\begin{pmatrix}\nu_L\\\ell_L\end{pmatrix} + (\bar{\nu}_L, \bar{\ell}_L)\begin{pmatrix}0\\1\end{pmatrix}\ell_R\right] & (64)\\ &= -\frac{G_\ell v}{\sqrt{2}}\bar{\ell}\ell - -\frac{G_\ell}{\sqrt{2}}\bar{\ell}\ell H & (65)\end{aligned}$$

with the difference, that their are no right-handed neutrinos in the SM. The first term represents the lepton mass (for the Higgs VEV $\langle\phi\rangle = \begin{pmatrix}0\\v/\sqrt{2}\end{pmatrix}$). The second term describes the additional interaction with the Higgs field $H \neq 0$.

1.7 Quantumchromodynamics

Until the 40ies of the 20th century, the description of matter was just based on the existence of protons and neutrons as elementary particles. In 1947, the charged pion π^- was discovered in cosmic radiation and shortly afterwards a new particle (the Kaon K^+) with a "strange" lifetime and a new quantum number, the *strangeness* had to be introduced.

In the 1950ies, a large variety of unstable particles could be detected in the first particle accelerators (*particle zoo*).

In 1964, M. GELL-MANN postulated the *quark* as fundamental fermion [16] to build all particles detected so far. In the same year, GREENBERG, HAN and NAMBU proposed the color quantum number for quarks and the force carrier of the strong force (*gluons*).

FRITZSCH, GELL-MANN and LEUTWYLER set up a Lagrangian for the Quantumchromodynamics (QCD) in 1973. It is given by:

$$\mathscr{L}_{\text{QCD}} = -\frac{1}{4}F^i_{\mu\nu}F^{\mu\nu,i} + \sum_{c=1}^{6}\bar{q}^i_\alpha(i\slashed{D}^\alpha_\beta)q^{\beta,i} \qquad (66)$$

where $\alpha,\beta = 0\ldots 2$ are color indices; the derivatives are summed over the quark flavors labeled by c. The QCD field strength tensor is given by

$$F^i_{\mu\nu} = \partial_\mu G^i_\nu - \partial_\nu G^i_\mu - g_s f_{ijk}G^j_\mu G^k_\nu \qquad (67)$$

with the strong coupling constant g_s and the structure constants f_{ijk} defined by

$$[\lambda_i, \lambda_j] = 2if_{ijk}\lambda_k \qquad (68)$$

with the *Gell-Mann matrices* λ. The covariant derivative in Eq. (66) is given by

$$D^\alpha_{\mu,\beta} = (D_\mu)_{\alpha\beta} = \partial_\mu \delta_{\alpha\beta} + ig_s G^i_\mu L^i_{\alpha\beta} \qquad (69)$$

with $L^i = \lambda^i/2$. The color interactions are diagonal in the flavor indices, but in general change the quark colors. They are purely vectorial, hence parity conserving. There are no bare mass terms for the quarks in Eq. (66). These would be allowed by QCD alone, but are forbidden by the chiral symmetry of the electroweak part of the theory, hence the quark masses are generated by spontaneous symmetry breaking. There are in addition effective ghost and gauge-fixing terms which enter into the quantization of both the $SU(3)$ and electroweak Lagrangians, and there is the possibility of adding an (unwanted) term which violates CP invariance [11].

1.8 Renormalization

Quantum field theories make predictions for observables depending on a finite number of parameters like fermion masses or coupling con-

1 THE STANDARD MODEL OF PARTICLE PHYSICS

$$\lambda^i = \begin{pmatrix} \tau^i & 0 \\ 0 & 0 \end{pmatrix} \quad i = 1, 2, 3$$

$$\lambda^4 = \begin{pmatrix} 0 & 0 & 1 \\ 0 & 0 & 0 \\ 1 & 0 & 0 \end{pmatrix} \qquad \lambda^5 = \begin{pmatrix} 0 & 0 & -i \\ 0 & 0 & 0 \\ i & 0 & 0 \end{pmatrix}$$

$$\lambda^6 = \begin{pmatrix} 0 & 0 & 0 \\ 0 & 0 & 1 \\ 0 & 1 & 0 \end{pmatrix} \qquad \lambda^7 = \begin{pmatrix} 0 & 0 & 0 \\ 0 & 0 & -i \\ 0 & i & 0 \end{pmatrix}$$

$$\lambda^8 = \frac{1}{\sqrt{3}} \begin{pmatrix} 1 & 0 & 0 \\ 0 & 1 & 0 \\ 0 & 0 & -2 \end{pmatrix}$$

Table 1: The Gell-Mann matrices.

stants. In general, those predictions are only valid for a restricted energy range. This range might be large, but is at least finite. Especially at certain points in phase space, these predictions start to differ strongly from physical reality, e. g. a space-time points where the propagators of inner lines exhibit poles or where physical effects such as electron self-energy become significant.

In order to burnish potentially dangerous regions in phase space and to improve the predictive power of a QFT, perturbation theory is applied by inserting additional orders of coupling constants (*fixed order calculation*). The predictions rendered by QFTs become more precise and the valid energy range larger the more orders of perturbation theory are added to the calculation.

Inserting higher powers of the coupling constants introduces a set of additional inner (*loops*) and external lines (*real emissions*) to the Feynman diagram of the process considered. Since the momentum flow through those additional diagrams is not restricted by the real physical process, ultraviolet divergences arise that would spoil any finite prediction of the theory. Within the development of the Standard

Model QFTs, methods had been worked out that drop such divergences.

The application of higher-order perturbation together with the cancellation of divergences (*counter-terms*) is called *renormalization*. If divergences can be canceled by introducing a finite number of additional, observable parameters, the theory is called renormalizable. As consequence, the bare Lagrangian $\mathscr{L}_0$ can be split into a renormalized Lagrangian $\mathscr{L}$ and a virtual action $\Delta\mathscr{L}$ containing the counter-terms [1][p. 226]:

$$\mathscr{L}_0 = \mathscr{L} + \Delta\mathscr{L}. \tag{70}$$

First developed for QED, it was shown that all theories with spontaneously broken symmetries contain ultraviolet divergences and are renormalizable [17][p. 347]. The reason is that the occurrence of higher order Feynman diagrams (loop diagrams and additional external lines) that yield energy dependent contributions to the physical coupling constants and parameters such as masses and the electric charge. QED, QCD and the GSW theory are all renormalizable theories.

Usually, the unphysical divergences that are finally absorbed in the physical parameters are denoted as Z_1, Z_2 and Z_3 (*renormalization constants*) where Z_1 contains the divergences of the vertex corrections (correction of the coupling constant) Z_2 contains the self-energy divergences in fermion propagators and Z_3 the divergences of the field propagators (photons, gluons). In the case of QED, the physical electrical charge e is obtained from the bare charge e_0 via

$$e = \frac{Z_2\sqrt{Z_3}}{Z_1}e_0. \tag{71}$$

Using the *Ward-Takahashi identities*, it can be shown that $Z_1 = Z_2$, so $e = \sqrt{Z_3}e_0$ [1][p. 232].

1.8.1 QCD and the Renormalization Group

Basically there is an infinite number of ways (*renormalization scheme*) to renormalize a QFT since the splitting of the bare Lagrangian $\mathscr{L}_0$ into the renormalized action $\mathscr{L}$ and the counter-terms contained in $\Delta\mathscr{L}$. As consequence, the renormalization constants $Z_{1...3}$ can be written as a function depending on the renormalization scheme R: $Z \to Z(R)$. Any renormalized quantity Γ can then be written as expression with the bare quantity Γ_0 with acting $Z(R)$ as operator on it:

$$\Gamma = Z(R)\Gamma_0. \qquad (72)$$

$Z(R)$ obeys all group laws including multiplication $Z(R'', R')Z(R', R) = Z(R'', R)$ and the identity $Z(R, R) = 1$. Since the variation of the renormalization energy scale μ does not change the unrenormalized quantity Γ_0, rewriting Γ_0 in terms of renormalized quantity Γ yields:

$$\frac{\partial}{\partial\mu}\Gamma_0 = 0 \qquad (73)$$

$$\Rightarrow 0 = \left(\mu\frac{\partial}{\partial\mu}Z^{-1/2}\right)\Gamma + Z^{-1/2}\left(\mu\frac{\partial}{\partial\mu}\Gamma\right). \qquad (74)$$

Together with the definitions

$$\beta(g) = \mu\frac{\partial g}{\partial\mu}, \qquad (75)$$

$$\gamma(g) = \mu\frac{\partial}{\partial\mu}\log\sqrt{Z}, \qquad (76)$$

$$m\gamma_m(g) = \mu\frac{\partial m}{\partial\mu}, \qquad (77)$$

Eq. (73) yields

$$\left(\mu\frac{\partial}{\partial\mu} + \beta(g)\frac{\partial}{\partial g} - n\gamma(g) + m\gamma_m(g)\frac{\partial}{\partial m}\right)\Gamma(p, g, m, \mu) = 0. \qquad (78)$$

(78) and (75) are called *renormalization group equations* [1][p. 478].

1.8.2 Asymptotic Freedom and Quark Confinement

Solving $Z_{1...3}$ for QCD, the renormalization group equations yield (at one-loop-level)

$$g^2(\mu) = \frac{g^2(\mu_0)}{1 + g^2(\mu_0)/8\pi^2 \left(\frac{11}{3}C_{ad} - \frac{4}{3}C_f\right)\log(\mu/\mu_0)} \quad (79)$$

as prescription for the strong coupling constant g (*running coupling*) [1][p. 484]. As consequence, the strong coupling constant decreases by increasing the considered energy scale, resulting in a free theory in the $\mu \to \infty$ limit. This is in contrast to QED (and also the the weak coupling), which exhibits an increasing coupling with increasing energy. Hence, quarks with a very high energy can be considered to be free Dirac particles whereas electrons are surrounded more and more by virtual photons and e^+e^- pairs.

1.9 The Standard Model

The Standard Model of Particles Physics combines QCD and the Yang-Mills theory of EW physics as a $SU(3) \times SU(2) \times U(1)$ symmetry group. It considers 6 leptons and 6 quarks where the leptons only are affected by the electroweak forces. The quarks additionally interact with the gluons as QCD force carriers.

The leptons are arranged in three left-handed isospin doublets

$$e_L = \begin{pmatrix} \nu_e \\ e \end{pmatrix}, \quad \mu_L = \begin{pmatrix} \nu_\mu \\ \mu \end{pmatrix}, \quad \tau_L = \begin{pmatrix} \nu_\tau \\ \tau \end{pmatrix}. \quad (80)$$

The charged leptons exist also as right-handed singlets e_R, μ_R and τ_R. The neutrinos do not exist as right-handed singlets since right-handed neutrinos could not be discovered experimentally.

1 THE STANDARD MODEL OF PARTICLE PHYSICS 25

The quark sector exist completely both as a group of left-handed doublets and right-handed singlets:

$$q_{1,L} = \begin{pmatrix} u \\ d \end{pmatrix}, \quad q_{2,L} = \begin{pmatrix} c \\ s \end{pmatrix}, \quad q_{3,L} = \begin{pmatrix} t \\ b \end{pmatrix}, \quad (81)$$

$$u_R, \quad c_R, \quad t_R,$$
$$d_R, \quad s_R, \quad b_R.$$

The lepton and the quark sector are distinguished in three families containing one (left-handed) lepton (quark) weak isospin doublet each. The lepton with $I_3 = +1/2$ is electrically charged whereas the lepton with $I_3 = -1/2$ (*neutrino*) is electrically neutral and does not interact with the electromagnetic force. While the charged leptons have a sizable mass ($M_\ell = 511 \, \text{keV} \ldots 1.7 \, \text{GeV}$), the neutrinos are nearly massless and within the theory of the Standard Model, are considered to be exactly massless[2]. Fig. 3 shows the leptons and quarks and their properties known in the SM.

Neutrons and protons are built of the lightest quarks (1st quark family) so the up and the down quark form all stable matter. Generally, quarks can combine to bounded states, the *hadrons*: *Mesons* consist of two quarks where one quark carries color and the second one the same anti-color, so in terms of color addition rules, a meson is "white". *Baryons* combine three quarks, the quarks carry the color charge red, green and blue. Again, the sum of all quarks colors is white.

---

[2]Meanwhile, neutrinos have been observed to mix by an oscillation and therefore must have a (very small) mass.

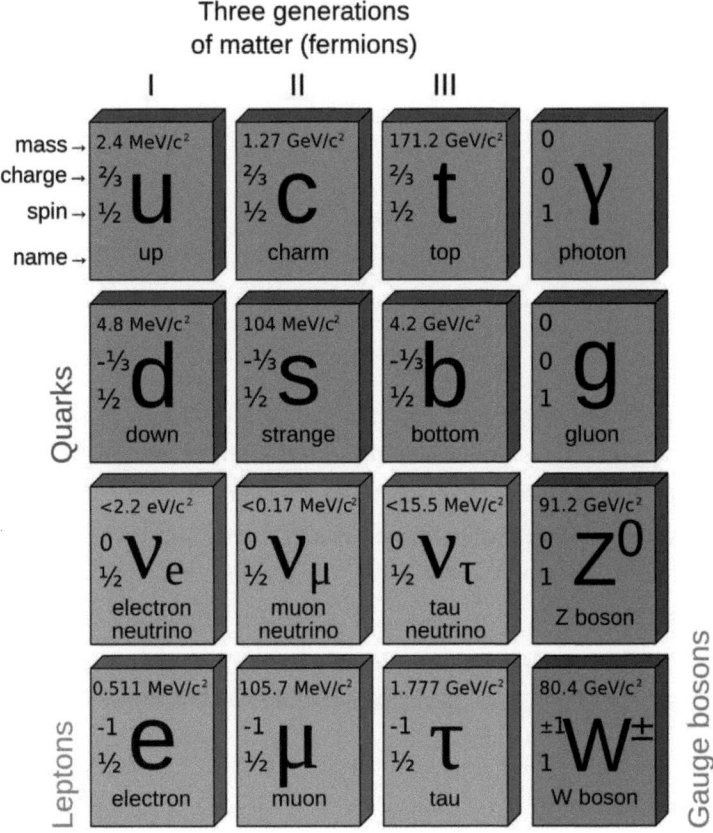

Figure 3: Fermions and gauge bosons of the Standard Model of Particles Physics.

1.10 Quark Mixing

In 1963, experiments showed that the change of strangeness ($|\Delta s| = 1$) is strongly suppressed. CABBIBO introduced a quark mixing matrix with a mixing angle to enable rare quark flavor transitions [18]:

$$\begin{pmatrix} |d'\rangle \\ |s'\rangle \end{pmatrix} = \begin{pmatrix} \cos\theta_C & \sin\theta_C \\ -\sin\theta_C & \cos\theta_C \end{pmatrix} \begin{pmatrix} |d\rangle \\ |s\rangle \end{pmatrix}. \qquad (82)$$

1 THE STANDARD MODEL OF PARTICLE PHYSICS

In 1973, KOBAYASHI and MASKAWA introduced a quark model with six flavors (three families) in the SM [19]. For this purpose, a 3×3 unitary matrix had to be introduced in order to translate the flavor eigenstates to mass eigenstates:

$$V_{\text{CKM}} = V_L^u V_L^{\dagger(d)} = \begin{pmatrix} V_{ud} & V_{us} & V_{ub} \\ V_{cd} & V_{cs} & V_{cb} \\ V_{td} & V_{ts} & V_{tb} \end{pmatrix}. \quad (83)$$

The quark flavor eigenstates transform to the mass eigenstates via

$$\begin{pmatrix} V_{ud} & V_{us} & V_{ub} \\ V_{cd} & V_{cs} & V_{cb} \\ V_{td} & V_{ts} & V_{tb} \end{pmatrix} \begin{pmatrix} |d\rangle \\ |s\rangle \\ |b\rangle \end{pmatrix} = \begin{pmatrix} |d'\rangle \\ |s'\rangle \\ |b'\rangle \end{pmatrix}. \quad (84)$$

In the standard representation, the CKM matrix introduces a complex phase δ that is responsible for CP violation [15]:

$$V_{\text{CKM}} = \begin{pmatrix} c_{12}c_{13} & s_{12}c_{13} & s_{13}e^{-i\delta} \\ -s_{12}c_{23} - c_{12}s_{23}s_{13}e^{i\delta} & c_{12}c_{23} - s_{12}s_{23}s_{13}e^{i\delta} & s_{23}c_{13} \\ s_{12}s_{23} - c_{12}c_{23}s_{13}e^{i\delta} & -c_{12}s_{23} - s_{12}c_{23}s_{13}e^{i\delta} & c_{23}c_{13} \end{pmatrix} \quad (85)$$

with $s_{ij} = \sin_{ij}$ and $c_{ij} = \cos_{ij}$. Since $s_{13} \ll s_{23} \ll s_{12}$, V_{CKM} can also be parametrized as

$$V_{\text{CKM}} = \begin{pmatrix} 1 - \lambda^2/2 & \lambda & A\lambda^3(\rho - i\eta) \\ -\lambda & 1 - \lambda^2/2 & A\lambda^2 \\ A\lambda^3(1 - \rho - i\eta) & -A\lambda^2 & 1 \end{pmatrix} + \mathcal{O}(\lambda^4) \quad (86)$$

a.k.a. the *Wolfenstein parametrization*. Eq. (85) can be translated to Eq. (86) via

$$s_{12} = \lambda = \frac{|V_{us}|}{\sqrt{|V_{ud}|^2 + |V_{us}|^2}}, \quad s_{23} = A\lambda^2 = \lambda \left| \frac{V_{cb}}{V_{us}} \right|,$$

$$s_{13}e^{i\delta} = V_{ub}^* = A\lambda^3(\rho + i\eta) = \frac{A\lambda^3(\bar\rho + i\bar\eta)\sqrt{1-A^2\lambda^4}}{\sqrt{1-\lambda^2}\left[1 - A^2\lambda^4(\bar\rho + i\bar\eta)\right]}.$$

Since V_{CKM} is assumed to be unitary, there should be six vanishing combinations since $\sum_i V_{ij}V_{ik}^* = \delta_{jk}$ and $\sum_j V_{ij}V_{kj}^* = \delta_{ik}$. These combinations can be illustrated as triangles in a complex plane, the *unitarity triangle*. Since V_{ij} are free parameters in the SM, a precise measurement is an important mean to show that there are only three quark families containing one up and one down-type quark each.

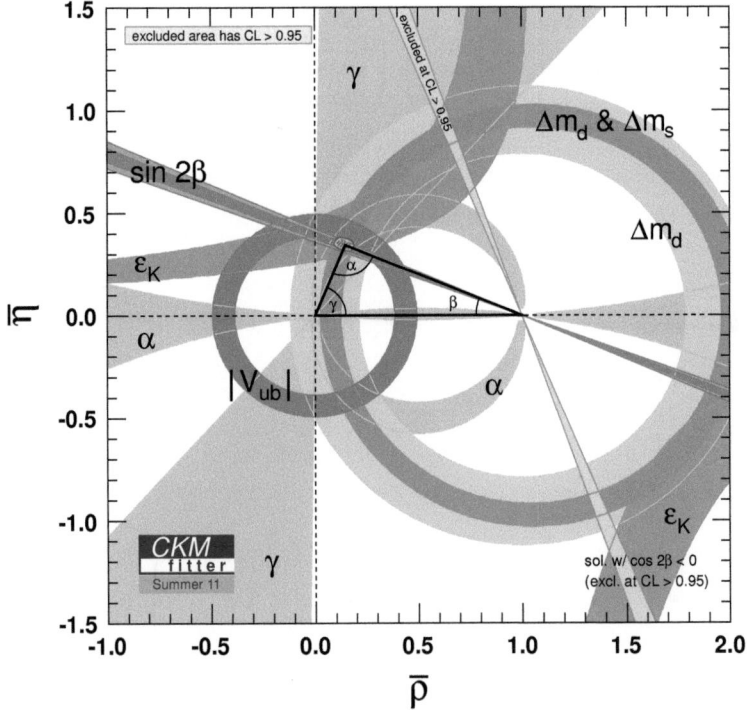

Figure 4: Unitarity triangle (latest result from the CKM fitter group [20]). The most commonly used $\bar\rho - \bar\eta$ representation is shown.

1.11 Challenges to the SM and Possible Extensions

Although the predictions of the Standard Model could be verified at a very high accuracy so far, there are still open questions it cannot answer; furthermore, there is a "philosophical" problem when extrapolating the predictions at the EW energy scale to the GUT[3] scale (*fine tuning problem*).
The main challenge on the long run will be the unification of the EW and strong force (GUT) and, beyond that, even the embedding of gravity in the Standard Model (*quantum gravity*). Nevertheless, any further unification beyond the weak and the EM force will be not accessible by experiments for a very long time[4]. Two very general and competitive approaches for describing quantum gravity are the concepts of string theory and quantum loop gravity.

1.11.1 The Hierarchy Problem

The hierarchy problem faces the question why the Planck energy scale so largely exceeds the EW scale. Scalar particles like the Higgs boson suffer from the fact, that within renormalization, the mass corrections of the Higgs bare mass can become very large (quadratic divergence of the corrected Higgs potential) [21]:

$$\lambda \int^\Lambda \frac{d^4}{k^2 - M_H^2} \quad \Rightarrow \quad \Delta V \propto \lambda \Lambda^2 \phi^\dagger \phi. \tag{87}$$

In order to bring the physical mass of the Higgs boson back to the expected range a careful (and unphysical) fine-tuning of $-\mu^2$ in Eq. (60) is required.[5]

---
[3]GUT = Great Unification Theory
[4]The energy scale for GUT is believed to be at the order of $\Lambda_{\text{GUT}} = \mathcal{O}(10^{16}\,\text{GeV})$
[5]Fermions do not suffer from quadratic divergences due to their chiral symmetry [21, p. 9]

A possible loophole is Super-Symmetry (SUSY), where additional, very heavy new particles introduce natural corrections of renormalization that cancel per default (see Sec. 1.11.3).

1.11.2 CP Violation

The *Sacharov criteria*, formulated by ANDREI SACHAROV in 1967, describe necessary conditions that had to exist during the Big Bang so that the Universe could evolve as we know it today. This includes also the requirement of a CP-violating mechanism in order to produce an excess of matter (and a deficit of anti-matter).

The SM predicts CP violating processes by introducing 1-loop-diagrams, but the predicted CP violation is by far too small to explain the matter excess we observe today.

1.11.3 Supersymmetry (SUSY)

Supersymmetry (SUSY) introduces a new sector of elementary particles that exist in parallel to the known leptons, quarks and gauge bosons. In SUSY, each fermion is assigned a bosonic partner and vice versa. The *Minimal Supersymmetric Standard Model* (MSSM) is of special interest, since it reduces the number of free parameters to a minimum of 120, where most of them are already excluded experimentally since they lead to very large flavor changing neutral currents or large electric dipole moments for the neutron and electron [21].

Furthermore, the MSSM imposes *R-parity* as additional conserved quantum number:

$$R = (-1)^{3B+L+2s}. \tag{88}$$

The R parity is positive for SM particles and negative for SUSY particles.

1.11.4 Kaluza-Klein Excitations

The Kaluza-Klein theory was the first theory that included gravity into the Standard Model. It predicts excited particle states within an additional fifth dimension that "sees" gravity larger than in the four space-time dimensions [22].

2 The Top Quark

The top quark is the 6th quark in the Standard Model an has an average mass (pole mass) of $(172.0\pm0.9\pm1.3)$ GeV [23] and is therefore by far the heaviest of the six quarks of the Standard Model of Particle Physics.

The top quark as second quark of the third left-handed quark doublet was discovered in 1995 by the CDF detector [24] and by DØ [25] at the Tevatron Laboratories in Michigan, USA. Fig. 5 shows the history of limit settings and direct measurements of the top quark mass during the last decades. It can be seen that already with indirect setting of EW limits, the mass of the top quark could be restricted quite successfully. The right plot of Fig. 5 summarizes the recent results of top quark mass measurements performed by the LHC detectors ATLAS and CMS.

The top quark is produced in hadron-hadron collisions predominantly via strong interaction. It decays nearly uniquely in the single mode $t \to W^+ b$ ($V_{tb} = 99.3\,\%$). The lifetime of the top quark is small compared to the timescale of hadronization [28, p. 4], so the top quark is the only quark that can be investigated as single quark. Rare decays and CP violation are very small regarding the SM predictions [28, p. 1] and such effects have neither been detected so far.

Since the mass of the top quark is nearby the energy scale of the electroweak (EW) symmetry breaking, the question arises why the top quark has a Yukawa coupling of the order $\mathcal{O}(1)$; it could play a very fundamental role in EW symmetry breaking: New physics would likely be manifested in anomalies of top quark production and decays [28, p. 1].

2 THE TOP QUARK

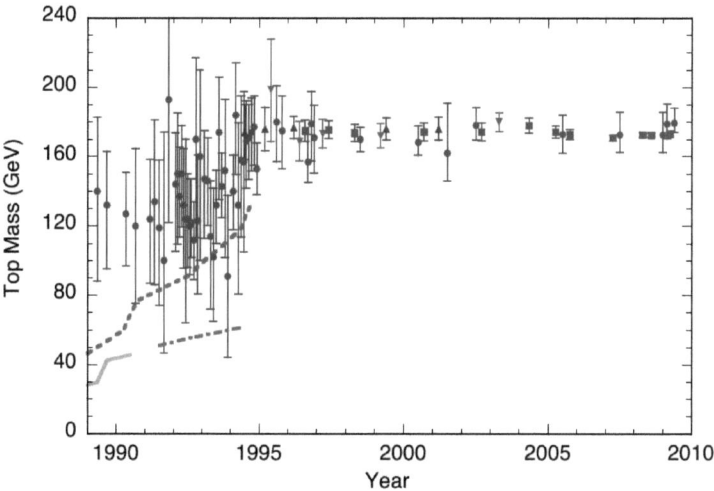

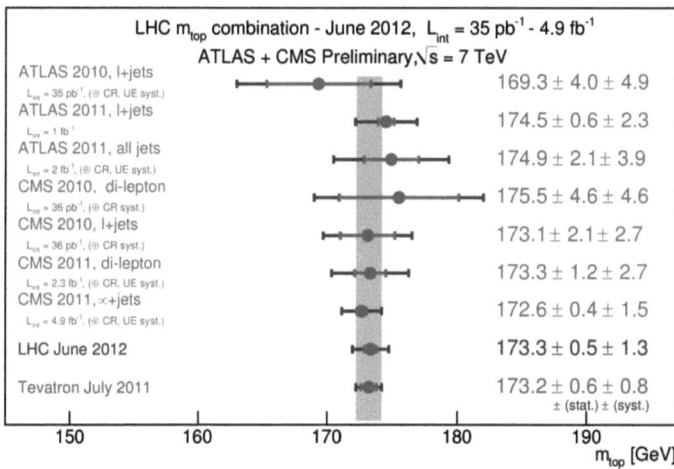

Figure 5: History of limit settings and direct measurements of the top quark during the last decades [26]. The green dots (●) show the indirect bounds from EW precision data. Triangles indicate direct measurements from CDF (▲) and DØ (▼); squares are world averages (■). The dashed lines are lower limits from direct searches by Spp̄S and the Tevatron, the solid line by e^+e^- colliders (LEP, PETRA). Recent LHC result for the top mass are shown in the right plot [27].

2.1 Properties of the Top Quark

The pole mass of the top quark is related to its Yukawa $y_t(\mu)$ coupling by

$$y_t(\mu) = 2^{3/4} G_F^{1/2} m_t (1 + \delta_t(\mu)) , \qquad (89)$$

where $\delta_t(\mu)$ is a radiative correction depending on the energy scale μ [28, p. 3]. Furthermore, the $\overline{\text{MS}}$ mass of the top quark $\overline{m}_t(\mu)$ can be defined taking into account radiative QCD corrections $\delta_{\text{QCD}}(\mu)$:

$$\overline{m}_t(\mu) = m_t (1 + \delta_{\text{QCD}}(\mu))^{-1} , \qquad (90)$$

assuming five massless quark flavors besides the top quark and a strong coupling constant of $\alpha_s = 0.03475$ considering a four-loop calculation [28, p. 3]. The on-shell decay width Γ_0 of the top quark is, assuming $|V_{tb}| = 1$:

$$\Gamma_0 = \frac{G_F m_t^3}{8\pi\sqrt{2}} = 1.76 \,\text{GeV} \qquad (91)$$

or, incorporating M_W as basis for a radiative correction at LO:

$$\Gamma_{\text{LO}}(t \to Wb)/|V_{tb}|^2 = \Gamma_0 \left(1 - 3\frac{M_W^4}{m_t^4} + 2\frac{M_W^6}{m_t^6}\right) = 0.885 \Gamma_0 = 1.56 \,\text{GeV} . \qquad (92)$$

There are a few other corrections (see Tab. 2), the final result for the corrected decay width is [28, p. 4]:

$$\Gamma(t \to Wb)/|V_{tb}|^2 \approx 0.807 \Gamma_0 = 1.42 \,\text{GeV} . \qquad (93)$$

2.1.1 Rôle of the Top Quark Mass in EW Precision Physics

The consistency of the SM strongly depends on the relation of parameters to each other, which are predicted by theory on the one hand and can be measured at high accuracy on the other hand. Models that extend the SM like the MSSM (see Sec. 1.11.3) change the theoretical

2 THE TOP QUARK

$M_W \neq 0$, correction at LO	-11.5%
α_s correction, $M_W = 0$	-9.5%
α_s and $M_W \neq 0$ correction	$+1.8\%$
α_s^2 correction, $M_W = 0$	-2.0%
α_s^2 and $M_W \neq 0$ correction	$+0.1\%$
EW correction	$+1.7\%$

Table 2: Corrections to the top quark mass, based on different radiative corrections and different assumptions of masses of other particles [28, p. 4].

expectations and can be tested indirectly.
Two types of uncertainties influence the significance of such model checks: Theoretical uncertainties which depend on the order of perturbation calculation (LO→NLO) and experimental uncertainties on the measured parameters. The top quark mass enters the EW precision observables as input parameter in loop corrections [28, p. 4]. Radiative corrections to the W mass M_W can be expressed by a general radiative correction Δr

$$M_W^2 = \frac{\frac{\pi \alpha}{\sqrt{2} G_F}}{\sin^2 \theta_w (1 - \Delta r)} \qquad (94)$$

with $\sin^2 \theta_w = 1 - \frac{M_W^2}{M_Z^2}$ [26]. The contribution from top quarks in the loop correction (see Fig. 6) is

$$(\Delta r)_{\text{top}} = \frac{3 G_F}{8\sqrt{2}\pi^2 \tan^2 \theta_w} m_t^2 \,. \qquad (95)$$

2.1.2 Electric Charge of the Top Quark

If the top quark is really the EW isospin partner of the bottom quark, it must have an electric charge of $Q_t = 2/3$. An alternative interpretation allows a right-handed, heavy-quark doublet $(Q_1, Q_4)_R$ where Q_1 could mix with the right-handed b-quark singlet b_R and, accordingly, Q_4 could mix with the SM top quark [29]. Since the electric charge of

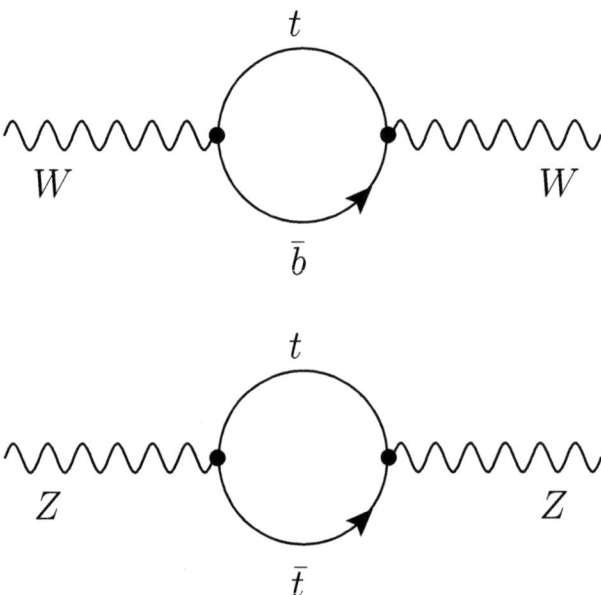

Figure 6: One-loop corrections contributing to M_Z and M_W.

Q_1 was $-1/3$, it could not distinguished from the SM bottom quark. The charge of Q_4 in this model was predicted to be $-4/3$ which would lead to a signature $Q_4 \to W^- b$.

Using the b-jet charge technique, the ratio ρ between the production of the SM top quark and Q_4 could be excluded at the 68 % CL for a exotic fraction of $\rho < 0.52$ and at 90 % CL for $\rho < 0.80$. The existence of Q_4 with the complete abundance of the SM top quark could be excluded at the 92.2 % CL [30].

Recent results published by the ATLAS collaboration could exclude the exotic $Q_t = -4/3$ scenario with a significance of more than 5σ [31].

In principle, the electric charge of the top quark can also be measured by testing the photon coupling to the top quark (see Sec. 3), which is only depending of the fermion charge in the SM $(-iQ_f^2 \bar{f} \gamma_\mu f)$. This method provides a much better experimental accuracy than the

b-charge method but suffers from a much lower cross section. This approach is not followed anymore because the exotic top quark scenario could already be excluded[6].

2.2 Top Quark Physics at Hadron Colliders

At hadron colliders, the colliding (anti-)protons are not point-like (Dirac) particles. They consist of three valence (anti-) quarks and numerous gluons that steadily produce virtual quark-anti-quark pairs (*sea quarks*). All these proton constituents carry a certain fraction of the overall proton momentum and energy, obeying a *parton density function* (PDF, see Sec. 2.2.1). As consequence, the CMS energy of the colliding quarks or gluons involved in the creation of the hard process is not a fixed value but differs from collision to collision. Hence, the cross section that is measured (*hadronic cross section*) σ has to be calculated as the convolution of the *partonic cross section* $\hat{\sigma}$ and the PDF $f(x, \mu^2)$ [26]

$$\sigma_{pp \to t\bar{t}} = \sum_{i,j=q,\bar{q},g} \int dx_i dx_j f_i(x_i, \mu_F^2) f_j(x_j, \mu_F^2) \hat{\sigma}(ij \to t\bar{t}, \hat{s}, \mu_R^2) \quad (96)$$

with the colliding partons i and j which can be quarks or gluons. The partonic cross sections $\hat{\sigma}(ij \to t\bar{t})$ have to be evaluated for all possible combinations of i and j, depending on the partonic CMS energy $\hat{s}$.

μ_F is the *factorization scale*, i.e. the energy scale on which the partons are assumed to be freely moving Dirac particles (*factorization theorem*). μ_R is the *renormalization scale*, i.e. the scale at which the matrix element for calculating $\hat{\sigma}$ is evaluated. The choice of both scales is arbitrary, a common setting is $\mu = \mu_F = \mu_R$ [26].

---

[6]Besides, not only the electric charge enters the $t\gamma$ coupling but also a hypothetical dipole moment. Thus, any measured deviation from the expected coupling would be somehow ambiguous (see Sec. 3.1).

2.2.1 Parton Model and Parton Distribution Functions (PDFs)

The cross section of an electron or positron interacting with a proton or anti-proton by the exchange of a photon can be expressed by a leptonic tensor $l_{\mu\nu}$ and a hadronic tensor $W_{\mu\nu}$. The hadronic tensor can be expressed by introducing two *form factors* W_1 and W_2 which depend on the momentum and energy transfer to the electron ($k \to k'$, $E \to E'$) only. Experimentally, it could be shown that W_1 and W_2 can be rewritten as two *structure functions*

$$F_1(x) := MW_1(q^2,\nu) \qquad F_2(x) := \nu W_2(q^2,\nu) \qquad (97)$$

called *Bjorken scaling* [1, pp. 461]. M is the proton mass, $q = k - k'$ the momentum transfer and $\nu = E - E'$ the energy transfer[7] to the electron respectively. x is defined as $x = -\frac{q^2}{2M\nu}$.
Assuming that a (anti)proton consists of a number of quarks and gluons (*parton model*) all of them carrying a fraction ξ of the whole (anti)proton, it can be shown that, by introducing an (unknown) function $f(\xi)$ that describes the distribution of the fraction of momenta of the partons, the structure functions can be put together in the *Callan-Gross relation* [1, p. 465]:

$$F_1(x) = \frac{1}{2}f(x), \qquad F_2(x) = xf(x) \quad \Rightarrow \quad 2xF_1(x) = F_2(x). \qquad (98)$$

Knowing all quark charges and the valence quark content of protons and exploiting baryon number conservation, $f(x)$ can be constrained for all quark types (*sum rules*) [32]:

$$\int_0^1 dx\,[u(x) + \bar{u}(x)] = 2 = 2 \cdot \int_0^1 dx\,[d(x) + \bar{d}(x)] \qquad (99)$$

---

[7]Generally, it is $\nu = \frac{p \cdot q}{M}$. $\nu = E - E'$ is valid due to the small electron mass.

and for sea quarks respectively:

$$\int_0^1 dx\, [q_S(x) + \bar{q}_S(x)] = 0 \qquad (q_S = c, s, b). \tag{100}$$

Assuming total energy-momentum conservation, the gluon distribution $g(x)$ function can be constrained either:

$$\int_0^1 dx\, x \left[\sum_{i=u,d,s,c,b} (q_i(x) + \bar{q}_i(x)) + g(x) \right] = 1. \tag{101}$$

Exploiting the DGLAP evolution equations [33, 34], the prescription for the scale evolution of the quark and gluon PDFs can be derived from the results of measurements determined at an energy scale μ:

$$\mu^2 \frac{\partial}{\partial \mu^2} f(x, \mu, \mu^2) = \sum_{j=q,g} \int_x^1 \frac{dy}{y} P_{ij}(\frac{x}{y}, \alpha_s(\mu^2)) f_j(y, \mu, \mu^2). \tag{102}$$

P_{ij} are called *evolution kernels*. The kernels can be obtained at any order of precision with fixed-order calculation via

$$\mu^2 \frac{\partial}{\partial \mu^2} f(x, \alpha_s(\mu)) = P_{ij}(x) + \mathcal{O}(\alpha_s^2). \tag{103}$$

The parton distributions can be measured in deep inelastic scattering processes. Fig. 7 shows the measured PDFs provided by the MSTW group.

2.3 Top Quark Production Mechanisms

Top quark production can be divided into two classes:

Single Top Quark Production Single top quarks are produced via weak interactions. There are three separate single top quark production processes which are of interest at the LHC [28, p. 38]:

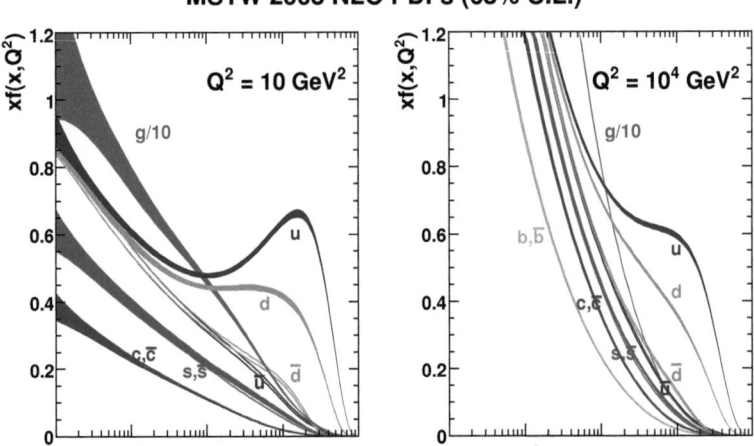

Figure 7: Illustration of the MSTW2008 PDF for all quark and anti-quark flavors and prediction for gluon density at NLO level with uncertainty bands at 68 % CL [33]. The evaluation for two different energy scales, $Q^2 = 10\,\text{GeV}^2$ (left plot) and $Q^2 = 10^4\,\text{GeV}^2$ (right plot) are shown.

- t-channel: In the t channel production, a space-like W boson ($q^2 \leq 0$) transfers momentum to a b sea quark from a proton and promotes it to a top quark (see Fig. 8).

- s-channel: A time-like W boson is produced in quark-anti-quark annihilation. The W boson decays into a top quark and an $\bar{b}$ quark ($W \to t\bar{b}$). The virtuality of the W boson becomes $q^2 \geq (m_t + m_b)^2$.

- Associated (or Wt) production: A virtual top quark is produced in weak interaction where a b quark is promoted to the virtual top quark and radiates an on-shell W boson ($q^2 = M_W^2$). The virtual top quark becomes on-shell by absorbing a gluon from the proton.

The t channel production mode has the largest cross section, namely about 1/3 of that of top quark pair production. The s-channel mode has the smallest cross section, with almost one order of magnitude less than the t-channel process. The cross section of the associated Wt production lies in between [28, p. 39].
There are several reasons to study single top quark processes at the LHC [28, p. 39]:

- The cross sections for single top quark production are proportional to $|V_{tb}|^2$, so V_{tb} can be measured directly.

- Single top quark events are background for other important processes (e.g. some Higgs decay channels).

- Measurement of the $V - A$ structure of the EW coupling of the top quark since single top quarks are produced with nearly 100 % polarization.

- New physics can be discernible in single top quark events by inducing non-SM weak interactions; either via higher order loop effects or by providing new production sources.

Top Quark Pair Production Top quark pairs ($pp \to t\bar{t}$) at hadron colliders are produced via strong interaction. A pair of top quarks ($t\bar{t}$) can be produced by gluon fusion processes and quark-anti-quark annihilation. Fig. 9 shows a set of representative Feynman diagrams that contribute to top quark pair production.
The partonic cross section for top quark pair production via quark-anti-quark annihilation $q\bar{q} \to t\bar{t}$ at LO is given by

$$\frac{d\hat{\sigma}}{d\hat{t}}(q\bar{q} \to t\bar{t}) = \frac{4\pi\alpha_s^2}{9\hat{s}^4} \left[(m_t^2 - \hat{t})^2 + (m_t^2 - \hat{u})^2 + 2m_t^2\hat{s}\right] \qquad (104)$$

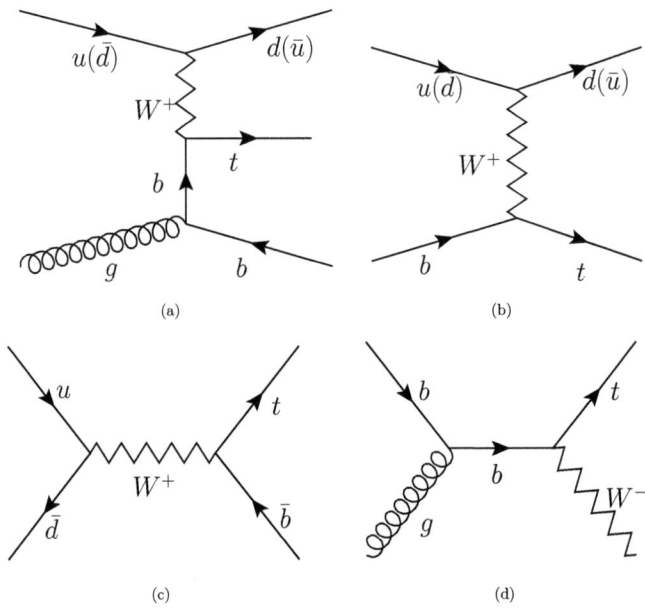

Figure 8: Feynman diagrams for single top quark production. (a) and (b) represent the t-channel, (c) the s-channel and (d) the associated production (Wt).

with the (partonic) Mandelstam variables $\hat{s} = (P_q + P_{\bar{q}})^2$, $\hat{t} = (P_q - P_t)^2$ and $\hat{u} = (P_q - P_t)^2$ [26]. For the $gg \to t\bar{t}$ production mode, the cross section reads [35]

$$\frac{d\hat{\sigma}}{d\hat{t}}(gg \to t\bar{t}) = \frac{4\pi\alpha_s^2}{\hat{s}^2}\left(\frac{1}{6\tau_1\tau_2} - \frac{3}{8}\right)\left[\frac{4m_t^2}{\hat{s}} + \tau_1^2 + \tau_2^2 - \frac{16m_t^4}{\tau_1\tau_2\hat{s}^2}\right] \quad (105)$$

with $\tau_1 = (m_t^2 - \hat{t})/\hat{s}$ and $\tau_1 = (m_t^2 - \hat{u})/\hat{s}$.

Fig. 10 shows the comparison of cross section measurements of top quark pair production for two CMS energies of 1.96 TeV (Tevatron) and 7 TeV (LHC) together with theoretical calculations at the NNLO level. The theoretical predictions for pp (LHC) and $p\bar{p}$ (Tevatron) begin to coincide with increasing $\sqrt{s}$, which happens for two reasons:

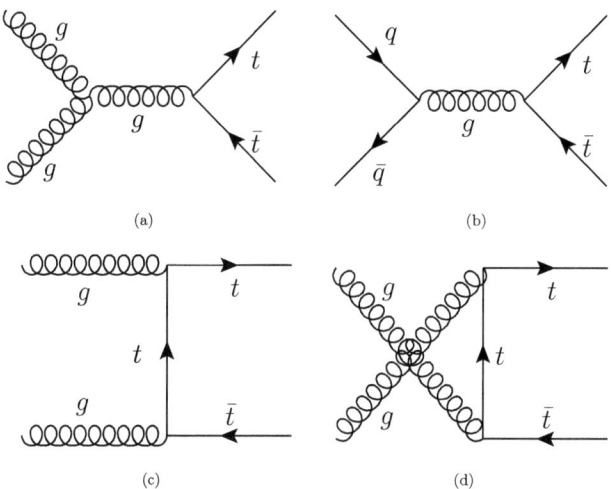

Figure 9: Feynman diagrams for top quark pair production. (a), (c) and (d) depict gluon fusion, (b) shows the process of quark-anti-quark annihilation. In the processes (c) and (d), two colliding gluons produce the top quark pair by the exchange of a highly virtual top quark.

- The gluon induced production mode becomes dominant whereas differences in the quark-anti-quark composition in protons and anti-protons are reduced with increasing momentum fractions $x_{i,j}$ for quarks q_i and q_j.

- Sea quarks, which are distributed equally in protons and anti-protons, are involved in quark-anti-quark annihilation to an increasing degree.

2.4 Top Quark Decay

Since the top quark decays nearly always into a b quark and a W boson, the event signature of top quark decays only depends on how the W decays. Each W boson can either decay leptonically ($W \to \ell\nu_\ell$, $\ell = e, \mu, \tau$) or hadronically, i.e. into two light quarks ($W \to q\bar{q}'$ with $q = u, c$ and $q' = d, s$). The decay of a W boson into a b quark and anything else is highly suppressed by the corresponding tiny entries of the CKM matrix.

Taking into account that the hadronic decay $W \to q_1 \bar{q}_2$ can occur in three color combinations and neglecting all quark and lepton masses, the BRs behave like

$$\text{BR}(W \to e\nu) : \text{BR}(W \to \mu\nu) : \text{BR}(W \to \tau\nu) : \text{BR}(W \to q_1 \bar{q}_2)$$
$$= 1 : 1 : 1 : 6 \qquad (106)$$

regarding the first two quark families ($q_1 = u, c$; $q_2 = d, s$). For the decay modes of a pair of top quarks, the combinatorics of all possible W boson decays can be classified as follows:

Di-Leptonic Decay Both W bosons decay into two leptons ($t\bar{t} \to \ell^+ \nu_\ell \ell^- \bar{\nu}_\ell b\bar{b}$, ($\ell = e, \mu, \tau$)). This decay channel is very clean since there are no hadronic jets except the ones formed by the two b quarks but has the smallest BR. Besides, it is difficult to perform a full kinematic reconstruction of the top quarks since the two neutrinos cannot be detected as separate particles. τ leptons are difficult to reconstruct so the only the combinations containing exclusively $W \to e\nu$ and $W \to \mu\nu$ are considered.

All Hadronic Decay Both W bosons decay into a quark-anti-quark pair, hence producing four jets plus two b-jets from the top quark decay ($t\bar{t} \to q_1 \bar{q}_2 q_3 \bar{q}_4 b\bar{b}$, ($q = u, d, c, s$)). This channel has the largest BR but suffers from a large combinatorial background because the jets are difficult to be assigned to the correct W boson. Besides, additional multijet production (underlying event) cannot be removed reliably.

Semi-Leptonic Decay One W boson decays into two leptons and the other one into a quark-anti-quark pair ($t\bar{t} \to q_1 \bar{q}_2 \ell \nu_\ell b\bar{b}$, ($\ell = e, \mu, \tau$, $q = u, d, c, s$)). This decay mode provides a trade-off between moderate combinatorial and multijet background and a quite large BR. Furthermore, the two top quarks can be fully reconstructed since there is only one neutrino which can be recovered by measuring the missing transverse energy of the event. Due to this trade-off, this decay mode is referred to as the "golden channel".

Tab. 3 lists the branching ratios of all possible combinations of the three decay modes. Excluding any occurrence of a τ lepton, a branching ratio of $\approx 30\,\%$ remains for the e+jets and μ+jets channel.

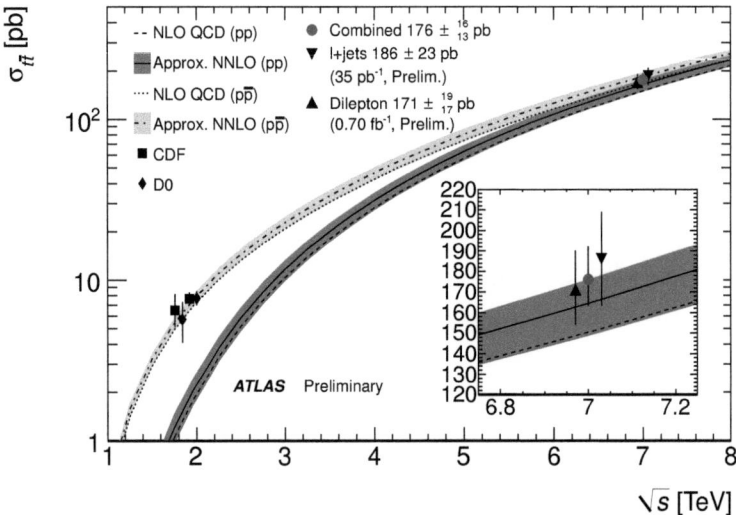

Figure 10: Experimental results for the cross section of top quark pair production at Tevatron (CDF, DØ) and the LHC (ATLAS, CMS) [36]. The measured values are in good agreement with the theoretical predictions.

Decay mode	BR
di-leptonic	
ee, $\mu\mu$, $\tau\tau$	1.2 %
$e\mu$, $e\tau$, $\mu\tau$	2.4 %
Sum	11.2 %
semi-leptonic	
(e, μ, τ)+jets	14.8 %
Sum	44.4 %
all hadronic	
Sum	44.4 %

Table 3: Branching ratios of all possible decay modes of top quark pair production.

3 Electroweak Couplings to the Top Quark

The top quark, like any other quark in the SM, should couple to photons and the Z boson as predicted. Since it decays before it hadronizes, the top quark is the only quark whose properties can be investigated as unconfined, bare quark directly.

The SM makes precise predictions on the structure and strength of EW couplings to charged particles. While the coupling of quarks to the W boson is simply $\propto \gamma^\mu(1-\gamma^5)$, the according vertices of photons and Z bosons have either a more complex $V-A$ structure ($c_A \neq c_V \neq 1$) or, in the case of photon coupling, are purely of vectorial nature. Deviations from the EW measurements or the occurrence of axial-vectorial or even tensorial contributions to the $t\gamma/tZ$ vertex would be either a hint for unknown particles or would raise suspicion that the top quark was not an elementary Dirac particle but had an internal structure (e. g. Technicolor) [37].

Deviations from the SM prediction of the $t\gamma$ or tZ coupling would become noticeable by a modified cross section, a modified momentum spectrum of the radiated γ/Z or a deviation in the angular distribution.

Since the cross section of top quark pair production with the associated emission of a photon is about a factor of ≈ 100 higher than the production with an associated Z boson, the measurement of the $t\bar{t}\gamma$ cross section is the first step towards the studies of gauge boson couplings to the top quark.

3.1 Approaches to Anomalous EW Top Quark Couplings

There are two approaches for describing additional, anomalous EW couplings to the top quark: The first one is induced by extending the γ^μ coupling to a general coupling Γ^μ that could contain any amount of vectorial, axial-vectorial and tensorial couplings [37]. The second one uses a generalized dim-6 operator approach and extends the SM Lagrangian by an effective Lagrangian [38]. Both approaches lead to equivalent results.

The general vertex for of the coupling of a top quark to a vector boson V can be written as

$$\Gamma_\mu^{tV}(k^2, q, \bar{q}) = -ie\left\{\gamma_\mu \left[F_{1V}^V(k^2) + \gamma_5 F_{1A}^V\right] \right. \\ \left. + \frac{\sigma_{\mu\nu}}{2m_t}(q+\bar{q})^\nu \left[iF_{2V}^V(k^2) + \gamma_5 F_{2A}^V(k^2)\right]\right\} \quad (107)$$

where tV describes the top quark-photon coupling $t\gamma$ as well as the coupling to Z bosons (tZ). This equation translates the vertex structure of tV to a general vectorial (F_{1V}), and axial-vectorial (F_{1A}) coupling. For on-shell (final state) photons, it is $F_{1A}^\gamma = 0$ [37, 38].

Furthermore, additional dipole moments might exist. In the case of $t\gamma$, F_{2V} can be related to a magnetic dipole moment (g_t), F_{2A} to an electric dipole moment (d_t^γ) of the top quark:

$$F_{2V}^\gamma(k^2 = 0) = Q_t \frac{g_t - 2}{2}, \quad F_{2A}^\gamma = \frac{2m_t}{e} d_t^\gamma. \quad (108)$$

In the case of non-vanishing dipole moments, F_{2V}^γ would deviate from the electric top quark charge and F_{2A}^γ would result in non-zero values. When considering the EW coupling of the top quark to the Z boson, F_{2V} and F_{2A} correspond to weak electric/magnetic dipole moments. In the SM, there are no tensorial couplings hence $F_{2V}^\gamma = F_{2V}^Z = 0$ and $F_{2A}^\gamma = F_{2A}^Z = 0$. The $t\gamma$ vertex coupling is determined by the electric charge of the top quark ($F_{1V}^\gamma = -\frac{2}{3}$) and completely of vectorial nature ($F_{1A}^\gamma = 0$). The SM tZ couplings obey the $V - A$ structure:

$$F_{1V}^Z = -\frac{1}{4\sin\theta_w \cos\theta_w}\left(1 - \frac{8}{3}\sin^2\theta_w\right), \quad F_{1A}^Z = \frac{1}{4\sin\theta_w \cos\theta_w}. \tag{109}$$

The form factors $F_{iV,A}^V$, ($i = 1, 2$) can be expanded from the SM values:

$$F_{iV,A}^V = F_{iV,A}^{V,\text{SM}} + \Delta F_{iV,A}^V(k^2) \tag{110}$$

with

$$\Delta F_{iV,A}^V(k^2) = \frac{\Delta F_{iV,A}^V(0)}{(1 + k^2/\Lambda_{FF}^2)^2} \quad (i = 1, 2) \tag{111}$$

where Λ_{FF} is the energy scale at which new physics would introduce anomalous EW couplings to the top quark. Current experimental results limit the tensorial couplings of the $t\gamma$ vertex to $-0.2 \leq F_{2V}^\gamma \leq 0.5$ and $|F_{2A}^\gamma| \leq 4.5$. The anomalous (axial-)vectorial $t\gamma$ couplings have not been constrained so far [37].

3.2 Top Quark Pair Production with an Additional Photon

As already mentioned, the process $pp \to t\bar{t}\gamma$ provides the largest cross section among all processes of top quark pair production with the associated generation of a (neutral) gauge boson. Hence, this process is the first one being considered at the LHC and subject of this thesis. As for any standard top quark pair analyses, the semi-leptonic decay mode provides a good trade-off between a high BR and a moderate S/B ratio for studies on top quark pair production with the associated radiation of a photon either. While the production and the decay of a $t\bar{t}$ pair can be separated since the top quark is so heavy, this is not possible anymore when an additional photon is generated in the process.

Since the photon is massless, it can be radiated from any electrically charged particle. Thus, considering final states with the top quarks and subsequent W bosons being decayed, Feynman diagrams with photon couplings to the top quark, b quark and the charged lepton and light quarks from the W decays will contribute to the full ME. It is not possible to uniquely distinguish between $t\gamma$ couplings and couplings to other particles due to interference terms, so the ME calculation has to be based on a full 7-particle final state (see Fig. 11):

$$q\bar{q}(gg) \to b\bar{b}\ell\nu j_1 j_2 \gamma. \tag{112}$$

Hence, a cross section measurement looking for a semi-leptonic $t\bar{t}$ decay signature plus an additional photon does not only include $t\gamma$ couplings.

Nevertheless, the measured phase space can be constrained in a way that it is enhanced by photon couplings to top quarks. For this purpose, the 7-particle process Eq. (112) can be divided into $q\bar{q}(gg) \to t\bar{t}\gamma$ (*radiative production*) and $t \to Wb\gamma$ (*radiative decay*) by linking top quark production and decay via a *narrow width approximation* [37]. The narrow width approximation reduces the Breit-Wigner distribution of the internal top quark propagators to a δ function:

$$\frac{1}{(s - m_t^2)^2 + m^2\Gamma^2} \to \frac{\pi}{m\Gamma}\delta(s - m_t^2). \tag{113}$$

Fig. 12 shows representative Feynman diagrams for radiative top quark production. Most of the photons are radiated from (virtual) top quarks, in quark-anti-quark annihilation processes, the photon can also couple to one of the incoming quarks. In radiative production, the top quark is on-shell after the photon has been emitted.

Accordingly, Fig. 13 shows representative Feynman diagrams for radiative top quark decay. Here, most of the photons will be radiated

from charged leptons, but also from quarks and the W boson. Here, the top quark is on-shell before it emits the photon.
The cross section of the process $t\bar{t}\gamma$ had been measured at the Tevatron collider for the first time and yields $\sigma_{t\bar{t}\gamma} = [0.18 \pm 0.07(\text{stat.}) \pm 0.04(\text{syst.}) \pm 0.01(\text{lumi.})]$ pb at a significance of $3\,\sigma$ [39].

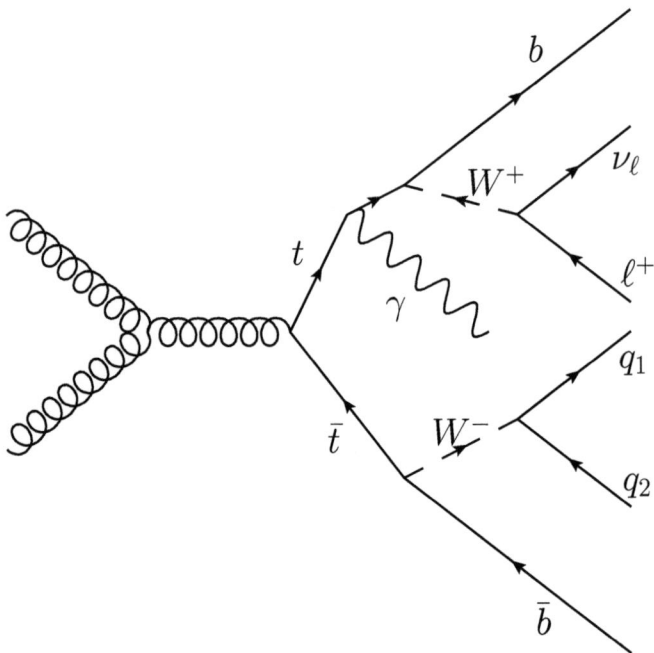

Figure 11: Representative Feynman diagram of the semi-leptonic $t\bar{t}$ decay with the emission of an additional photon (7-particle final state).

3 ELECTROWEAK COUPLINGS
TO THE TOP QUARK 53

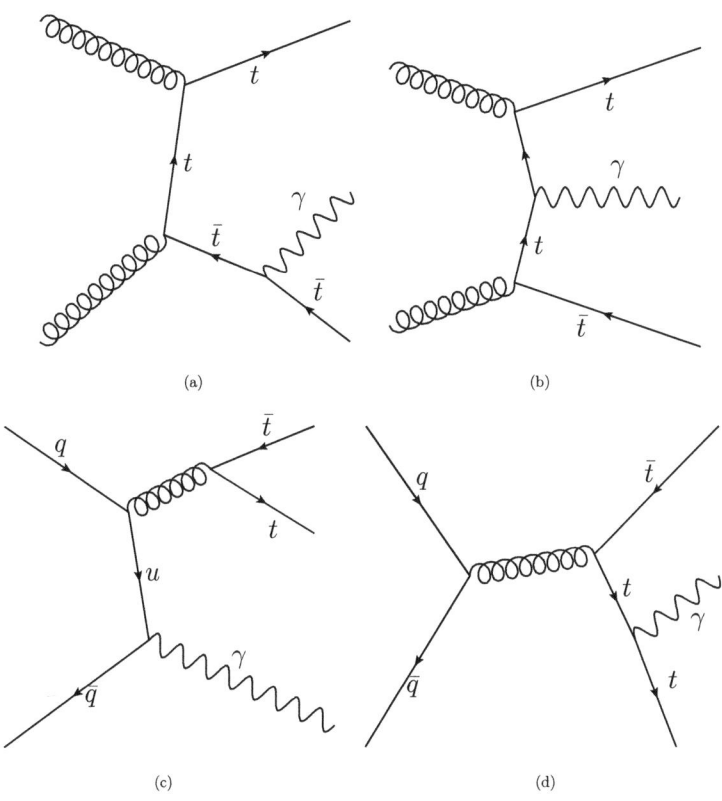

Figure 12: Representative Feynman diagrams of radiative top quark production. While in (a), (c) and (d) the photon couples to a (virtual) top quark, it can also be emitted from an incoming quark (b), which is not sensitive to the $t\gamma$ coupling.

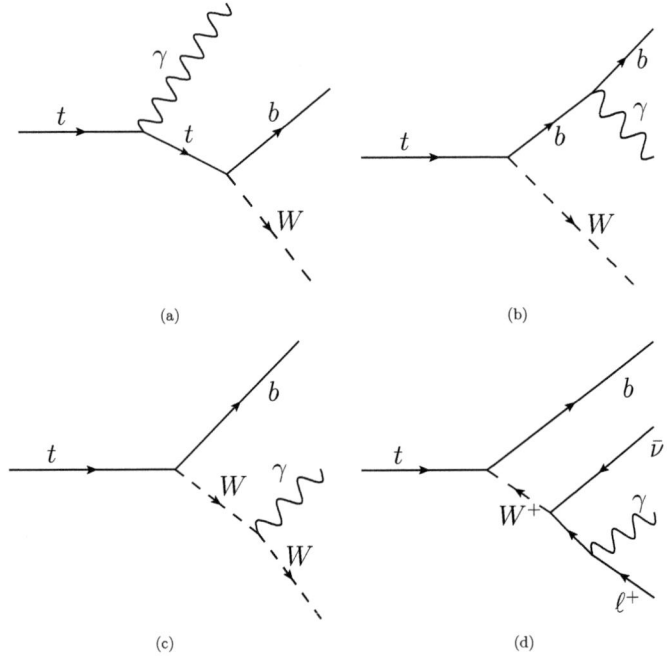

Figure 13: Representative Feynman diagrams of radiative top quark decay. The only possible $t\gamma$ coupling is shown in (a). In most cases, the photon is emitted by quark (b), the W boson (c) or a charged lepton (d).

4 The Large Hadron Collider

The Large Hadron Collider (LHC) is a proton-proton hadron collider built in a 26.7 km long ring tunnel which extends across the Swiss/French border. The tunnel reaches depths down from 45 m to 170 m below ground level and had already been used for the Large Electron Positron Collider (LEP) which was removed from the tunnel in the year 2000. Due to geological reasons, the tunnel plane is tilted by 1.4° towards the Lake Geneva [40, p. 3].

Two beam tubes with proton beams circulating in opposite directions are combined with the cooling and magnet systems in one housing (see Fig. 14). The LHC ring consists of eight straight segments and eight arc segments. Each straight segment is about 528 m long. The arcs are divided into 23 cells, each of them containing three dipole bending magnets [41, p. 25]. The vacuum in the beam tubes has to be kept in the range between 10^{-10} and 10^{-11} mbar [41, p. 339].

The protons are accelerated by RF cavities that operate at a frequency of 400 MHz. The accelerating voltage per beam is 16 MV [41, p. 132]. The beams are kept on their circular path by 1232 superconducting dipole magnets made of a Nb-Ti alloy, operated at a temperature of 1.9 K, cooled by super-fluid helium [41, p. 155, p. 161]. The magnetic field strength ranges from 0.54 T at the 450 GeV SPS injection energy up to 8.33 T at the design beam energy of 7 TeV [41, p. 164]. Superconducting quadrupole, sextupole and octupole magnets keep the proton beams in focused orbits. A schematic cross section through a dipole magnet is shown in Fig. 16.

The protons are first pre-accelerated by a linear accelerator (LINAC) and two small storage rings (BOOSTER and PS) sequentially and then injected into the SPS, where the protons are kept ready at an energy of 450 GeV for injection into the LHC. Fig. 15 shows a schematic overview over all LHC components and further experiments attached to them.

Figure 14: View into the LHC tunnel. The blue tube is one of the dipole segments [42].

4 THE LARGE HADRON COLLIDER

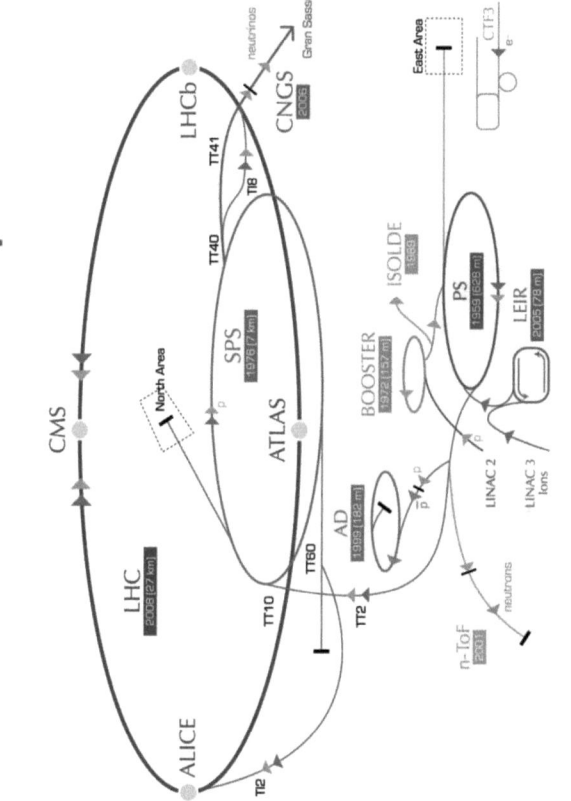

Figure 15: Schematic overview of the LHC, its detectors and pre-accelerators [43].

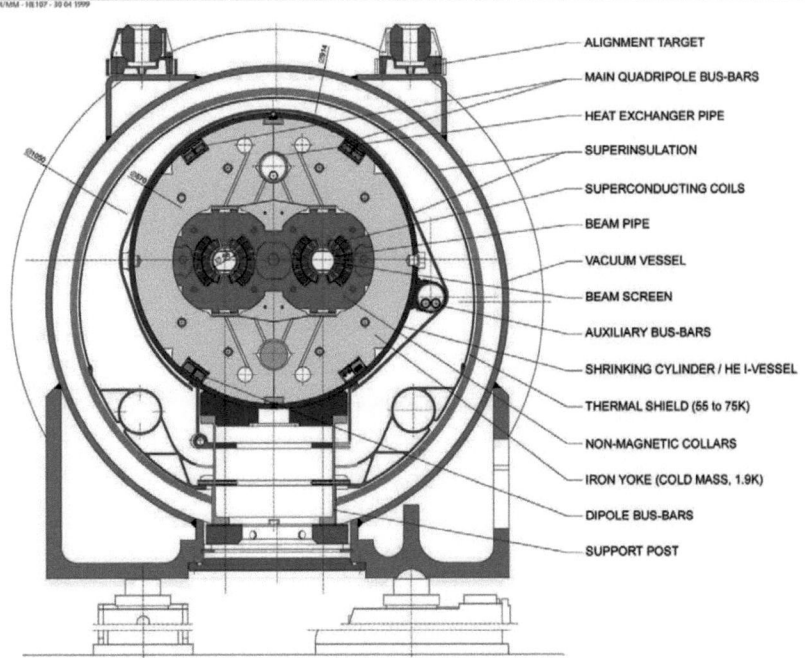

Figure 16: Sketch of a LHC dipole in cross section [44].

4.1 Proton Pre-Acceleration

The protons are first accelerated by the LINAC linear accelerator which produces a proton current of 180 mA with a hydrogen consumption of 4 ml/h [45, p. 13]. The pulse length of the proton beam is $> 100\,\mu s \cdot c$.

The proton pulses are transferred to the BOOSTER, a first ring storage accelerator that brings the protons to an energy of 1.4 GeV [45, p. 17].

When injected into the PS ring, the proton pulses are split into 4 batches of 72 bunches with a bunch spacing of 25 ns, increasing the proton energy to 25 GeV. Alternative bunch spacings of 75 ns and 50 ns are possible [45, pp. 48]. The PS can deliver its bunch batches every 3.6 s [45, p. 17]. After the acceleration, each bunch has a length of 11 ns $\cdot c$. This length is reduced to $\approx 4\,\text{ns} \cdot c$ by an adiabatic bunch rotation [45, p. 48].

Finally, the protons are accelerated to 450 GeV in the SPS, the final storage ring before the injection of the bunches into the LHC [45, p. 81]. The bunches a ramped to 450 GeV after up to four PS injection cycles of the 3.6 s PS intervals have been injected at a ramping speed of typically 78 GeV/s. The combination of bunch injections from the PS and the subsequent ramping in the SPS is called *super-cycle*. 24 super-cycles are required to fill both photon rings of the LHC.

4.2 LHC Performance

The LHC is designed to operate at a maximum of 2808 bunches per beam at an interval of 25 ns [41, p. 21]. The number of events N_{events} produced by the LHC per time interval is given by

$$\frac{dN_{\text{event}}}{dt} = L \cdot \sigma_{\text{event}} \tag{114}$$

with the *machine luminosity* L [41, p. 21]. L is only depending on machine parameters and follows a Gaussian law which describes the beam diameter:

$$L = \frac{N_b^2 n_b f_{\text{rev}} \gamma_r}{4\pi \epsilon_n \beta^*} F . \tag{115}$$

N_b is the number of particles per bunch, f_{rev} the revolution frequency, n_b the number of bunches per beam, ϵ_n the normalized beam emittance, γ_r the relativistic γ factor, β^* the β-function at the collision point and F the geometric luminosity reduction factor due to the crossing angle at the interaction point (IP):

$$F = \sqrt{1 + \left(\frac{\theta_c \sigma_z}{2\sigma^*}\right)^2}^{-1} . \tag{116}$$

θ_c is the full crossing angle at the IP, σ_z the RMS of the bunch length and σ^* the RMS of the transverse beam diameter at the IP. The design luminosity of the LHC for the ATLAS and the CMS detector is estimated to be $L = 10^{34} \, \text{cm}^{-2} \, \text{s}^{-1}$. For a CMS energy of $\sqrt{s} = 8 \, \text{TeV}$, an instantaneous luminosity of approximately $8 \cdot 10^{33} \, \text{cm}^{-2} \, \text{s}^{-1}$ has been reached by the end of September in 2012 which is just below this limit. Due to this high luminosity aimed for, $p\bar{p}$ collisions are not feasible since anti-protons cannot be delivered at a sufficient rate; two proton beams are used instead. Two counter-rotating proton beams require separated opposite magnetic dipole fields and separated vacuum chambers.

The particle density in one bunch is limited by the non-linear increase of number of interactions in beam-beam collisions which is given by

$$\xi = \frac{N_{\text{bunch}} r_p}{4\pi \epsilon_n} \tag{117}$$

where $r_p = e^2/(4\pi \epsilon_0 m_p c^2)$ is the classical proton radius. Due to machine safety issues, a minimum aperture of 10σ in units of the beam

4 THE LARGE HADRON COLLIDER

size is required. For the maximum achievable β^*, a beam emittance of $\epsilon_r = 3.75\,\text{m}$ is expected, together with the maximum ξ leading to a maximum bunch size of $N_\text{bunch} = 1.15 \cdot 10^{11}$.

A third limitation of L is the maximum magnetic dipole field which restricts the maximum achievable γ_r. At its design CMS energy of 14 TeV, a magnetic dipole field strength of 8.33 T is required.

4.2.1 Beam Lifetime

Due to beam emittance and degradation of particle density (beam collisions), the instantaneous machine luminosity is not constant over time [41, p. 23]. The decay time constant due to losses of particles from beam collisions is given by

$$\tau_\text{nucl.} = \frac{N_\text{tot,0}}{L\sigma_\text{tot}k} \qquad (118)$$

where $N_\text{tot,0}$ is the initial beam density, σ_tot the total cross section ($10^{-25}\,\text{cm}^{-2}$ at $\sqrt{s} = 14\,\text{TeV}$) and k the number of IPs. Assuming the design luminosity of $L = 10^{34}\,\text{cm}^{-2}\,\text{s}^{-1}$, $\tau_\text{nucl.} = 44.85\,\text{h}$ is estimated. The time dependency of the luminosity and bunch density is

$$N_\text{tot}(t) = \frac{N_\text{tot,0}}{1 + t/\tau_\text{nucl.}} \quad \Rightarrow \quad L(t) = \frac{L_0}{(1 + t/\tau_\text{nucl.})^2}. \qquad (119)$$

The time relevant for the beam lifetime is $\tau_{1/e}$, i.e. the time until the bunch intensity has decreased to $1/e$ of its initial value:

$$t_{1/e} = (\sqrt{e - 1})\tau \qquad (120)$$

which yields a decay time of $\tau_{1/e} = 14.9\,\text{h}$ including additional losses [41, p. 106, p. 24]. With each turn, 6.7 keV of stored proton energy is lost due to synchrotron radiation [41, p. 108].

4.3 Luminosity Measurement

The measurement of the luminosity is important to obtain a cross section from a number of detected events (see Eq. (114)). The general method for calibrating the ATLAS luminosity scale remains based on *van der Meer scans* (vdM scan). The vdM scan assumes that the counting rate of events is proportional to a vertical displacement of the colliding beams w.r.t to each other [46].

There are two primary detectors used to make bunch-by-bunch luminosity measurements: LUCID and BCM [47].

LUCID is a Cerenkov detector specifically designed for measuring the luminosity in ATLAS. Sixteen aluminum tubes filled with C_4F_{10} gas surround the beam-pipe on each side of the interaction point (IP) at a distance of 17 m, covering the pseudo-rapidity range $5.6 < |\eta| < 6.0$. The Cerenkov photons created by charged particles in the gas are reflected by the tube walls and detected by photo-multipliers (PMT) situated at the back end of the tubes.

The Beam Conditions Monitor (BCM) consists of four small diamond sensors on each side of the ATLAS IP arranged around the beam-pipe in a cross pattern.

4.4 The LHC Detectors

The LHC ring provides four interaction points, each one equipped with a detector designed for specific purposes. The two largest ones, CMS and ATLAS, are general-purpose "discovery machines". Their design differs in a way to enable more or less independent measurements of the same physical processes thus being able to significantly increase the power of discovery. A detailed description of the ATLAS detector is given in Sec. 5.

One central question that the SM cannot answer so far is how the universe could create a sufficient excess of matter compared to the amount of antimatter shortly after the Big Bang. One mechanism that can cause such an excess is the CP violation (see Sec. 1.11.2), which is predicted by the SM to some extent but at an insufficient magnitude. The LHCb detector is dedicated to answer this question by examining the CP violation in flavor transitions of B mesons.

ALICE is a detector designed to investigate the properties of the *quark gluon plasma*, a state of matter that existed in the very early universe before it had cooled down so far that quarks could hadronize to protons and neutrons. To operate ALICE, the LHC is filled with heavy ions (typically Pb) for a few weeks during the operating period in one year.

4.5 Pile-up

While operating on design luminosity and a bunch spacing of 25 ns, in average $\langle n_{BX} \rangle = 23$ bunch crossings per beam circulation can be expected in the ATLAS detector [48]. Since the cross section of "interesting" physical processes is magnitudes below the cross section of background processes which represent the dominant fraction of the collisions, the probability that two processes of interest will happen within the same bunch crossing is very small.

Nevertheless, additional particles will be detected that are not part of the considered proton-proton interaction. This phenomenon is called *pile-up*. There are two different kinds of pile-up processes:

In-Time Pile-up (Event Pile-up) Within each bunch crossing, there is the possibility that more than one proton pair reacts and produces a collision signature in the detector. Assuming that the number of pp collisions per bunch follows a Poissonian distribution as well as the number of particles produced per *pp* collision, the average number of particles produced per bunch crossing is

$$N(pp) = \sum_{n=0}^{\infty} P_{n_{mb}}(n) \cdot e^{-n \cdot N_p} \frac{(n \cdot N_p)^N}{N!}. \qquad (121)$$

Out-Time Pile-up In some of the liquid argon calorimeters, the drift time for signals across the argon gap is ≥ 400 ns [48]. This means that signals from more than 20 bunch crossings are superimposed. In order to limit the effects of detector pileup and to yield a fast response for energy measurements in calorimeter cells, a fast shaped pulse is derived from the drift current signal.

Even with that fast signal shaping, some noise background remains due to superimposed drift signals and is given by

$$S = \sum_j E_j g(t_j) \qquad (122)$$

when an energy of E_j is deposited in one calorimeter cell in the j-th bunch crossing. $g(t_j)$ is the pulse shape at the time t_j. In average, due to alternating polarizations of $g(t_j)$, there should be no mean noise:

$$\langle S \rangle = \langle E \rangle \sum_j g(t_j) \stackrel{\langle g \rangle = 0}{=} 0. \qquad (123)$$

Due to statistical fluctuations, an event-by-event pileup background noise can be expected:

$$\sigma^2(S) = \sigma^2(E) \sum_j g^2(t_j). \qquad (124)$$

4.5.1 Pile-up Suppression

One of the ways to reduce pile-up contributions to the overall detector noise is using topological clustering to preferentially select cells with

4 THE LARGE HADRON COLLIDER 65

a high signal to noise ratio [48]. In calculating σ_{noise}, the expected contribution of pile-up noise to the total cell noise is calculated from Monte Carlo simulations. The values for the noise are dependent on the luminosity, specified by the number of minimum bias collisions per bunch crossing, and are calculated by a software tool. This procedure introduces a bias in the mean pile-up E_T.

5 The ATLAS Detector

The ATLAS detector is, from the dimensional point of view, the largest one of the four main LHC detectors. Like CMS, it has been designed as a multi-purpose "discovery machine" and allows for competitive discoveries and measurements. A CAD sketch of the ATLAS detector is shown in Fig. 17.

As the particle interactions take place in the CMS frame, the ATLAS detector is built in an onion skin shaped manner around the interaction point. The placement of the components of colliding beam detectors is generally built in the following order (from inside to outside):

1. a *tracking system* to reconstruct the tracks of charged particles,

2. an *electro-magnetic calorimeter* (EMC) to determine the energy of electrons and photons,

3. an *hadronic calorimeter* (HC) for the energy measurement of hadrons (jets),

4. a *muon spectrometer* to reconstruct the momentum of muons.

Besides, three magnet systems (one central solenoid in the middle, eight toroids located on the outside and two end-cap solenoids) are installed to curve the tracks of charged particles in order to be able to measure their momenta. Fig. 18 illustrates how several types of particles are detected and can be distinguished from each other.

5.1 Coordinate System

The z-direction of the ATLAS detector is determined by the beam direction, the x-y plane is the plane transverse to the beam direction where the positive x-direction points from the interaction point to the center of the LHC ring. The positive y-axis points upwards. The

5 THE ATLAS DETECTOR

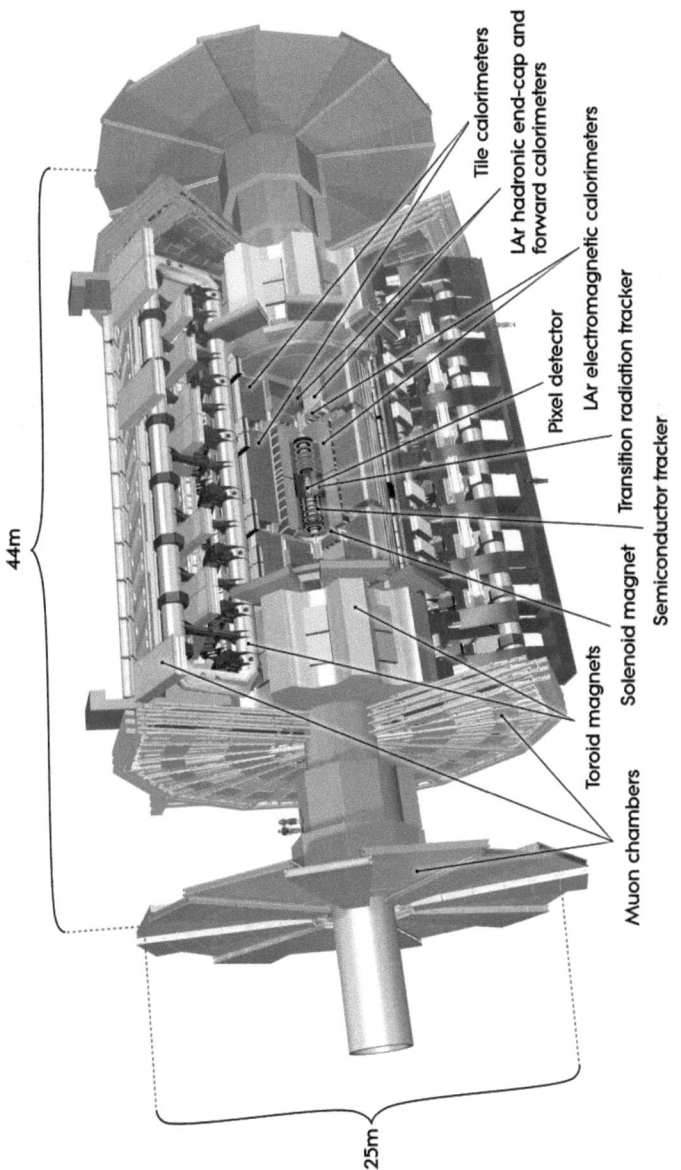

Figure 17: The ATLAS detector (CAD generated image) [49]

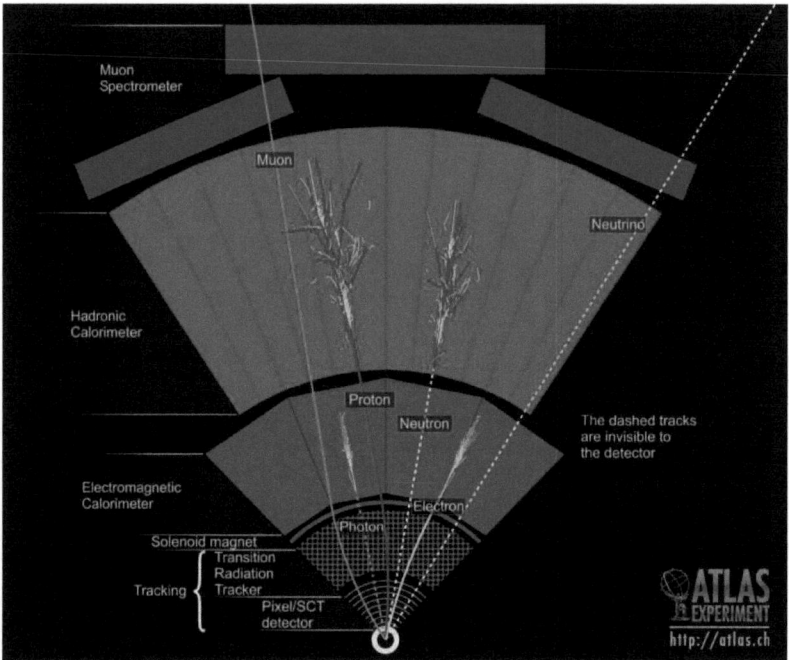

Figure 18: Particle detection with the ATLAS detector [50]. Charged particles, like charged leptons, muons and hadrons leave track information in the inner detector, whereas neutral particles (photons, neutrons) do not. Hadrons transverse the EMC and are detected in the HC additionally. Electrons and photons only deposit a small fraction of their energy in the HC.

azimuthal angle ϕ is measured perpendicularly around the beam axis; the polar angle θ is measured from the beam axis [51, p. 3]. The polar angle is usually expressed in terms of the *pseudo-rapidity* $\eta := -\ln\tan\frac{\theta}{2}$. The angular distance between objects is given by $\Delta R := \sqrt{\Delta\eta^2 + \Delta\phi^2}$.

Fig. 19 illustrates the specification of the ATLAS coordinate system.

5.2 The Inner Detector

The purpose of the inner detector (ID) is to measure the tracks of charged particles and to determine their production vertex. The usage

5 THE ATLAS DETECTOR

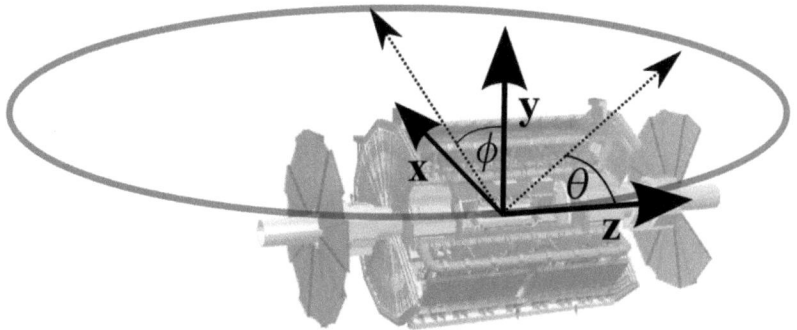

Figure 19: Coordinate system of the ATLAS detector.

of a highly granulated pixel detector as innermost layer, followed by the semi-conductive strip detector (SCT) and a straw tube tracker as 3rd layer, acting as *transition radiation tracker* (TRT), build a trade-off between high resolution and moderate costs.

By crossing at least three pixel layers and four SCT layers, together with typically 36 tracking points provided by the TRT, a robust track reconstruction is possible [52, p. 5].

The inner detector cavity has a length of 7 m and an outer radius of 115 cm and consists of three parts: One barrel for particle tracking for $|\eta| < 1$ and two identical end-caps for particle detection in the forward/backward direction. Within the barrel part, approx. 80 cm of the available radius are effectively used for detectors. [52, p. 5].

Fig. 20 shows the assembly of the components of the complete inner detector in a 3D CAD view.

5.2.1 The Pixel Detector

The ATLAS pixel detector consists of three concentric layers in the barrel part and three layers for each end-cap region.

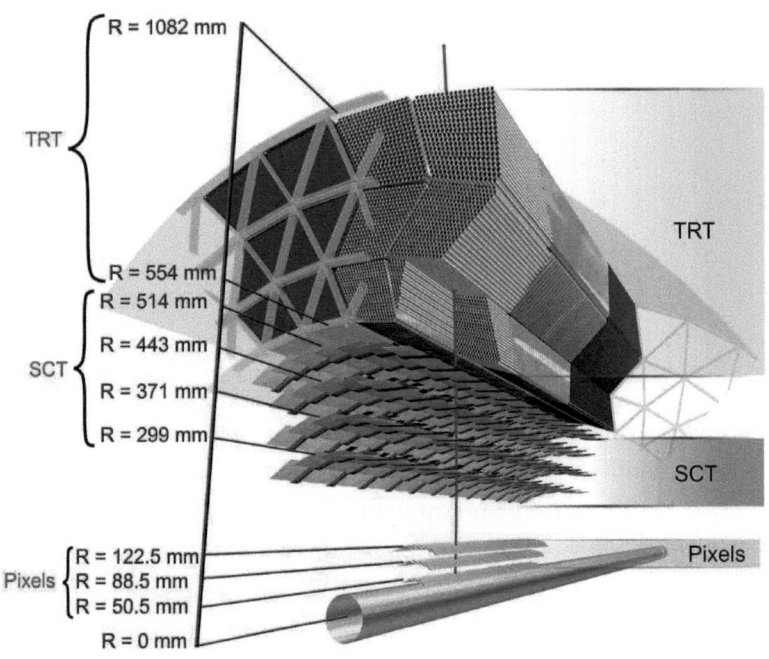

Figure 20: The inner detector (barrel part) as artistic 3D overview. The inner section is the pixel detector with three layers, positioned at radii of 50.0 mm, 88.5 mm and 122.5 mm. The pixel detector is followed by the SCT which is made up of four layers at $R = 299$ mm, 371 mm, 443 mm and 514 mm. The TRT as the outermost subsystem contains three tube layers between $R = 554$ mm and $R = 1082$ mm [53].

The innermost barrel layer (also called "b layer"[8]) installed at radii of 4 cm, 11 cm and 14 cm w.r.t. the beam axis. The end-cap parts of the pixel detector consist of four disks each. The barrel and the end-cap parts are made up from almost identical modules where 1500 modules are assembled in the barrel layers and 1000 modules in the

---

[8]The denotation "b layer" has physical reasons. Long-living B mesons decay after a finite decay length, the resulting, displaced decay vertex is reconstructed mostly by this layer.

disks [52, p. 8]. The individual pixels are each 300 µm long and 50 µm wide [54, p. 147].
The pixel detector provides typical spatial resolutions of $\sigma(R\varphi) = 12$ µm among the R-ϕ plane; in the barrel layers, the resolution is $\sigma(z) = 66$ µm in the z-direction and $\sigma(R) = 77$ µm considering the R-direction of the disks respectively [54, p. 18].

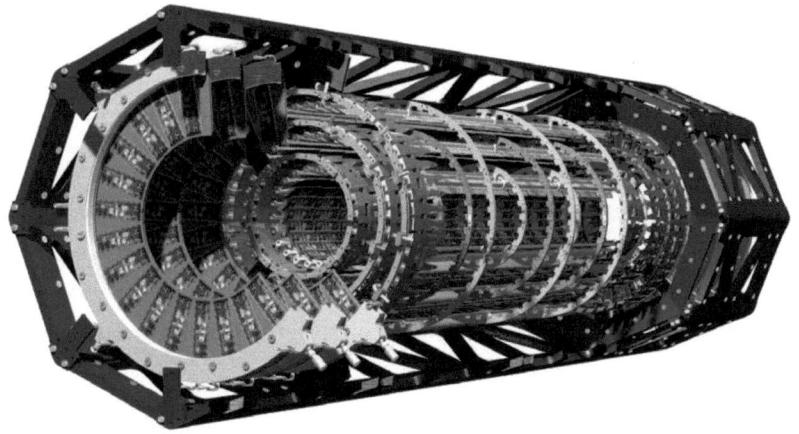

Figure 21: Schematic sketch of the pixel detector. The three end-cap modules on both sides and the three barrel layers can be seen [55].

Each barrel module has the dimensions 62.4 mm × 22.4 mm and is equipped with 61440 pixel elements. The pixels are read out by 16 chips, each of them responsible for an array of 24 × 160 pixels. The modules are overlapping on the support structure in order to give hermetic coverage [52, p. 8].

The read-out electronics of each pixel module group the data in pairs of columns in order to increase data bandwidth [54, p. 65]. Hence data is drained pair-wise, empty channels are skipped. The signal of the pixels is digitized and transmitted via optical fibers (two transmission and one reception fiber).

Fig. 21 shows a schematic overview of the pixel detector.

5.2.2 The Silicon Strip Detector

The silicon strip detector (a.k.a. semi-conductor tracker (SCT)) is built around the pixel detector and is designed to provide additional four precision measurements per track in the intermediate radial range, contributing to the measurement of momentum, impact parameter and vertex position, as well as providing good pattern recognition by the use of high granularity [52, p. 8].
The barrel SCT uses four layers of silicon micro-strip detectors, each one with the dimension $6.36 \times 6.40 \, \text{cm}^2$ and 768 readout strips. [52, p. 8].
Each module consists of four detectors. On each side of the module, two detectors are wire-bonded together to form 12.8 cm long strips [52, p. 8]. The SCT contains $61 \, \text{m}^2$ of silicon detectors, with 6.2 millions readout channels. The spatial resolution is $16 \, \mu\text{m}$ in the $R - \varphi$ and $580 \, \mu\text{m}$ in the z-direction. Tracks can be distinguished if they are separated by more than $\approx 200 \, \mu\text{m}$. [52, p. 8].
The SCT barrel modules are mounted on local supports and on carbon-fiber cylinders which carry the cooling system; the four complete barrels are situated at radii of 300 mm, 373 mm, 447 mm and 520 mm (see Fig. 20). The forward modules are grouped in three rings assembled on nine wheels. [52, p. 8].

5.2.3 The Transition Radiation Tracker (TRT)

The transition radiation tracker (TRT) is based on the usage of *straw detectors*. Straws are thin drift chamber tubes filled with inert gas (the ATLAS TRT uses Xenon). The TRT provides typically 36 additional track points for track reconstruction.
Each straw is 4 mm in diameter with a maximum straw length of 150 cm. The barrel contains about 50,000 straws, each divided into two halves at the center in order to reduce the occupancy and read-out at each end. The end-caps contain 320,000 radial straws. The total

number of electronic channels is 420,000.
Each channel provides a drift-time measurement, giving a spatial resolution of 170 µm per straw. [52, p. 9].
The barrel section is built of individual modules with 329 to 793 axial straws each. Each of the two end-caps consists of 18 wheels. The 14 wheels nearest to the interaction point cover a radial range from 64 to 103 cm, while the last four wheels extend to an inner radius of 48 cm [52, p. 10].

5.3 The Calorimetry System

The ATLAS calorimetry system consists of an electromagnetic calorimeter (EMC) covering the pseudo-rapidity region $|\eta| < 3.2$, a hadronic barrel calorimeter, covering $|\eta| < 1.7$, hadronic end-cap calorimeters covering $1.5 < |\eta| < 3.2$ and forward calorimeters covering $3.1 < |\eta| < 4.9$ [51, p. 11], as shown in Fig. 22.

The EMC is a lead/liquid argon (LAr) detector with an accordion geometry (Fig. 23). Over the pseudo-rapidity range $|\eta| < 1.8$, it is preceded by a pre-sampler detector which is used to correct for the energy loss in the material (ID, cryostats, coil) upstream of the calorimeter [51, p. 12].

The hadronic barrel calorimeter (HBC) is a cylinder divided into three sections: the central barrel and two identical extended barrels. It is based on a sampling technique with plastic scintillator plates (tiles) embedded in an iron absorber [51, p. 12].

The barrel EM calorimeter is contained in a barrel cryostat, which surrounds the ID cavity. The solenoid which supplies the 2 T magnetic field to the ID is integrated into the vacuum of the barrel cryostat and is placed in front of the EM calorimeter. Two end-cap cryostats house the end-cap EMC and hadronic calorimeters, as well as the integrated forward calorimeter [51, p. 12].

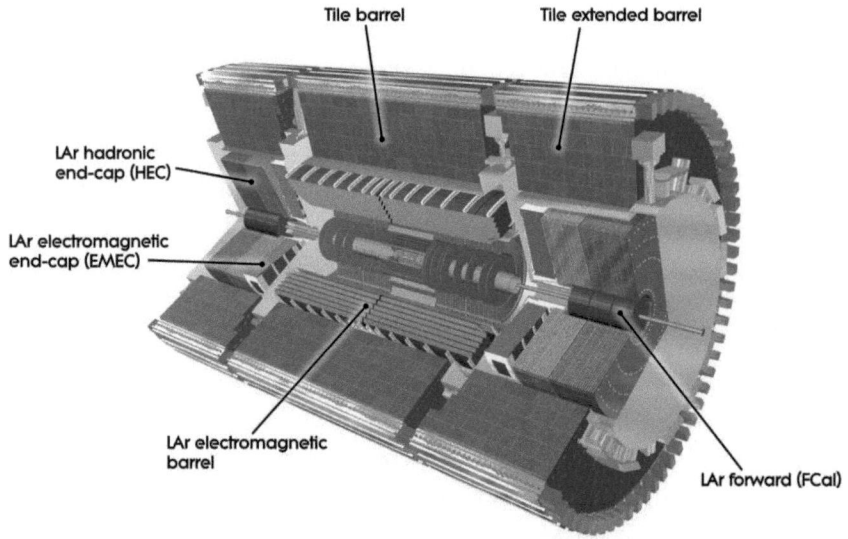

Figure 22: The ATLAS calorimetry system as schematic overview. The EMC is drawn in golden color, the surrounding HC in green/blue colors [56].

5.3.1 The Electromagnetic Calorimeter

The EMC is divided into a barrel part, covering a pseudo-rapidity range of $|\eta| < 1.475$ and two end-caps ($1.375 < |\eta| < 3.2$). The barrel calorimeter consists of two identical half-barrels, separated by a small gap (6 mm) at $z = 0$. Each end-cap calorimeter is mechanically divided into two coaxial wheels: an outer wheel covering the region $1.375 < |\eta| < 2.5$, and an inner wheel covering the region $2.5 < |\eta| < 3.2$.

The EMC is a lead LAr detector with lead absorbers designed in an accordion shape and is segmented into three longitudinal sections [51, p. 12]. The strip section is equipped with narrow strips with a pitch of $\approx 4\,\mathrm{mm}$ in the η direction. This section enhances particle identification ($\gamma \leftrightarrow \pi^0$, $e \leftrightarrow \pi$ separation, etc.). The middle section is transver-

5 THE ATLAS DETECTOR

sally segmented into square towers of size $\Delta\eta \times \Delta\phi = 0.025 \times 0.025$ (see Fig. 24). Fig. 23 shows a part of the EMC with the typical accordion structure.

The total calorimeter thickness up to the end of the second section is $\approx 24X_0$, the total number of read-out channels is $\approx 190,000$ [51, pp. 13].

Figure 23: Photograph of the EM LAr calorimeter. The accordion structure is clearly visible [57].

5.3.2 The Hadronic Calorimeter

The ATLAS hadronic calorimeters cover the pseudo-rapidity range of $|\eta| < 4.9$ using different techniques. For $|\eta| < 1.7$, the iron scintillating tile technique is used for the barrel and extended barrel tile calorimeters and for partially instrumenting the gap between them with the intermediate tile calorimeter (ITC). Over the range of $\approx 1.5 < |\eta| < 4.9$, LAr calorimeters have been chosen: the hadronic end-cap calorimeter

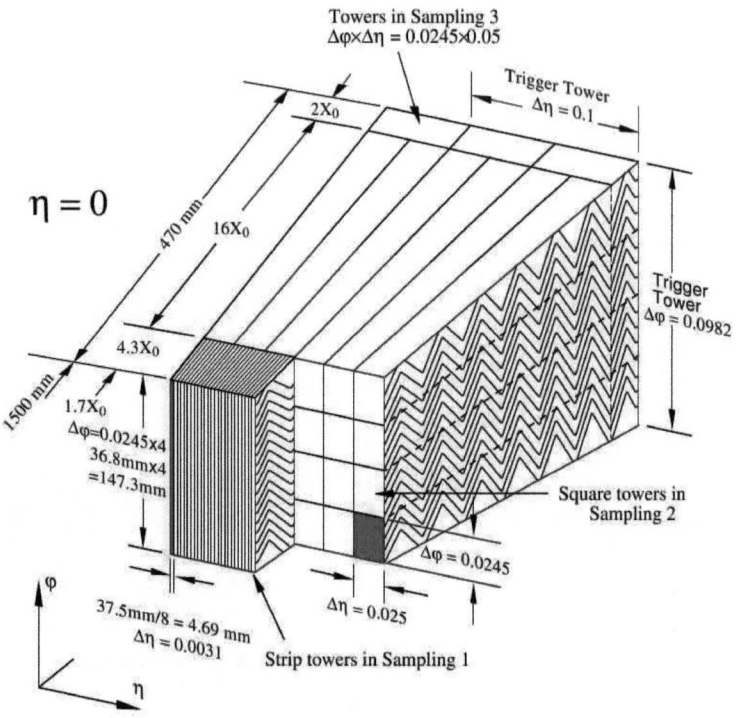

Figure 24: Granularity of the three layers of the LAr EMC [58].

(HEC) extends to $|\eta| < 3.2$, while the range $3.1 < |\eta| < 4.9$ is covered by the high-density forward calorimeter (FCAL). [51, pp. 14].

Tile Calorimeter The large hadronic barrel calorimeter is a sampling calorimeter using iron as absorber material and scintillating tiles as the active material, the scintillation photons are transmitted via wavelength shifting (WLS) fibers to two separate photo-multipliers (PMTs).
The tile calorimeter is composed of one barrel and two extended barrels. Radially, the tile calorimeter extends from an inner radius of

2.28 m to an outer radius of 4.25 m. It is longitudinally segmented in three layers. Azimuthally, the barrel and extended barrels are divided into 64 modules. The granularity is $\Delta\eta \times \Delta\phi = 0.1 \times 0.1$ (0.2×0.1 in the last layer). The total number of channels is about 10,000. The calorimeter is placed behind the EMC and the solenoid coil.
The barrel cylinder covers the region $|\eta| < 1.0$, the extended barrel the region $0.8 < |\eta| < 1.7$. [51, p. 15].

LAr Hadronic End-Cap Calorimeters   Each HEC consists of two independent wheels of an outer radius of 2.03 m. The upstream wheel is built from 25 mm copper plates, whereas the one farther from the interaction point uses 50 mm plates. [51, p. 15].
Each wheel is built out of 32 identical modules and a central ring and is divided into two longitudinal segments. The weight of the first (second) wheel is 67 (90) tons.
The end-cap EMC reaches $|\eta| = 3.2$, thereby overlapping the forward calorimeter [51, p. 16].

LAr Forward Calorimeter   The FCAL is integrated into the end-cap cryostat, with a front face at about 4.7 m from the interaction point and consists of three sections. In each section the calorimeter consists of a metal matrix with regularly spaced longitudinal channels filled with concentric rods and tubes with LAr in the gap between rods and tubes as sensitive medium. [51, p. 16].

5.4   The Muon Spectrometer

The ATLAS muon spectrometer is based on the magnetic deflection of muon tracks in the large superconducting toroid magnets, instrumented with separate trigger and high-precision tracking chambers.

Over the range $|\eta| \leq 1.0$, magnetic bending is provided by the barrel toroid. For $1.4 \leq |\eta| \leq 2.7$, muon tracks are bent by two smaller end-cap magnets inserted into both ends of the barrel toroid. Over a pseudo-rapidity range of $1.0 \leq |\eta| \leq 1.4$ (transition region), magnetic deflection is provided by a combination of the barrel and end-cap fields [51, p. 17].

In the barrel region, tracks are measured in chambers arranged in three cylindrical layers around the beam axis; in the transition and end-cap regions, the chambers are installed vertically, also in three stations. Over most of the η range, the measurement of the track coordinates is provided by Monitored Drift Tubes (MDTs). At large pseudo-rapidities and close to the interaction point, Cathode Strip Chambers (CSCs) with higher granularity are used in the innermost plane over $2.0 < |\eta| < 2.7$.

The precision measurement of the muon tracks is performed in the $R - z$ projection, parallel to the bending direction of the magnetic field; the axial coordinate z is measured in the barrel and the radial coordinate R in the transition and end-cap regions.

5.4.1 Monitored drift tubes (MDT)

The basic detection elements of the MDT chambers are aluminum tubes of 30 mm diameter and 400 µm wall thickness with a 50 µm diameter central Wolfram-Rhenium wire. The tubes are operated with a mixture of 93% Ar and 7% CO_2. The single wire resolution is ≈ 80 µm [51, p. 20].

5.4.2 Cathode Strip Chambers (CSC)

The CSCs are designed as multi-wire proportional chambers with cathode strip read-out and with a symmetric cell in which the anode-cathode spacing is equal to the anode wire pitch. The precision coordinate is obtained by measuring the charge induced on the segmented

cathode by the electron avalanche formed on the anode wire.
The cathode strips for the precision measurement are placed orthogonal to the anode wires. The anode wire pitch is 2.54 mm, the cathode read-out pitch is 5.08 mm. A measurement of the transverse coordinate is obtained from orthogonal strips oriented in parallel to the anode wires, which form the second cathode of the chamber. The CSCs are arranged in 2 × 4 layers [51, p. 21].

5.4.3 Resistive Plate Chambers (RPC)

The RPC is a gaseous detector providing a typical space-time resolution of 1 cm × 1 ns with digital read-out. The basic RPC unit is a narrow gas gap formed by two parallel resistive plates. The primary ionization electrons are multiplied to avalanches by a strong electric field [51, p. 21].

A trigger chamber is made from two rectangular detector layers, each one read out by two orthogonal series of pick-up strips: the "η strips" are aligned parallel to the MDT wires and provide the bending view of the trigger detector; the "ϕ strips", being orthogonal to the MDT wires, provide the second coordinate measurement which is also required for the off-line pattern recognition.

5.4.4 Thin Gap Chambers (TGC)

The TGCs are similar in design compared to the CSCs, with the difference that the anode wire pitch is larger than the cathode-anode distance. Signals from the anode wires, arranged parallel to the MDT wires, provide the trigger information together with read-out strips arranged orthogonal to the wires. These read-out strips are also used to measure the second coordinate [51, p. 22].

The main dimensional characteristics of the TGCs are a cathode-cathode distance (gas gap) of 2.8 mm, a wire pitch of 1.8 mm, and a wire diameter of 50 µm. [51, p. 22].

5.5 The Magnet Systems

In order to determine the momenta of charged particles, the ATLAS detector is equipped with three magnet systems:

- one **central solenoid**,
- one **barrel toroid**,
- two **end-cap toroids**.

5.5.1 The Central Solenoid

The Central Solenoid is a superconducting solenoid designed to provide a magnetic field of 2 T in the central tracking volume [59].
The solenoid is 5.3 m long with a bore of 2.4 m, has a thickness of 45 mm, a weight of almost 6 t and operates at a current of 7,600 A.

5.5.2 The Barrel Toroid

The barrel toroid consists of eight coils, each of them built by two windings with the dimensions 25 m long and 5 m wide. It generates the magnetic field for the central region of the muon detector. The coils are grouped in a torus shape maintained by a system of 16 supporting rings. Each coil is mounted in an individual cryostat and indirectly cooled by liquid helium. One single coil has a weight of some 40 t to which its cryostat adds another 40 t. The total barrel toroid assembly has an outer diameter of 20 m and approaches a weight of 830 t. The toroid is operated at a current of 20,500 A and reaches a peak field of 4 T. A photograph of the central toroid system is presented in Fig. 25.

5 THE ATLAS DETECTOR

Figure 25: View into the ATLAS detector with the 8 barrel toroid coils at the outer border [60].

5.5.3 The End-Cap Toroids

Two end-cap toroids are positioned inside the Barrel Toroid, one at each end of the central solenoid. They provide the magnetic field in the forward regions of the ATLAS detector across a radial span of 1.7 to 5.0 m. The eight coils of each end-cap toroid are assembled as a single unit inside one large cryostat.

Each coil has a size of about $4 \times 4.5\,\text{m}^2$ and a weight of roughly 13 t. One complete end-cap toroid has a diameter of 11 m, a width of 5 m and a weight of almost 240 t. The end-cap toroids are electrically connected in series with the barrel toroid and likewise operated at a current of 20,500 A, providing a peak field of 4 T. Fig. 26 shows a photograph of the end-cap toroid before its installation in the detector.

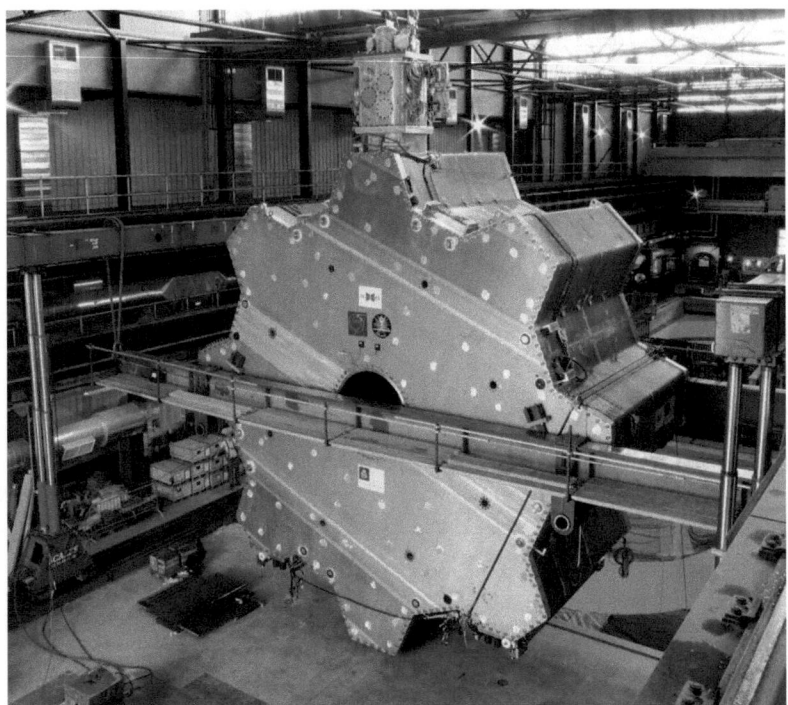

Figure 26: The ATLAS end-cap toroid before its insertion into the detector [61].

5.6 The Trigger and Data Acquisition (DAQ) System

The ATLAS trigger and data acquisition (DAQ) system is based on three levels of on-line event selection. Each trigger level refines the decisions made at the previous level and, where necessary, applies additional selection criteria. Starting from an initial design bunch crossing rate of 40 MHz (corresponds to an interaction rate of $\approx 10^9$ Hz at a design luminosity of $10^{34}\,\text{cm}^{-2}\,\text{s}^{-1}$), the rate of selected events must be reduced to ≈ 100 Hz for permanent storage. This requires an overall rejection factor of 10^7 against "minimum bias" events [51, p.

5 THE ATLAS DETECTOR

25]. Fig. 27 shows a schematic overview of the ATLAS trigger and DAQ system.

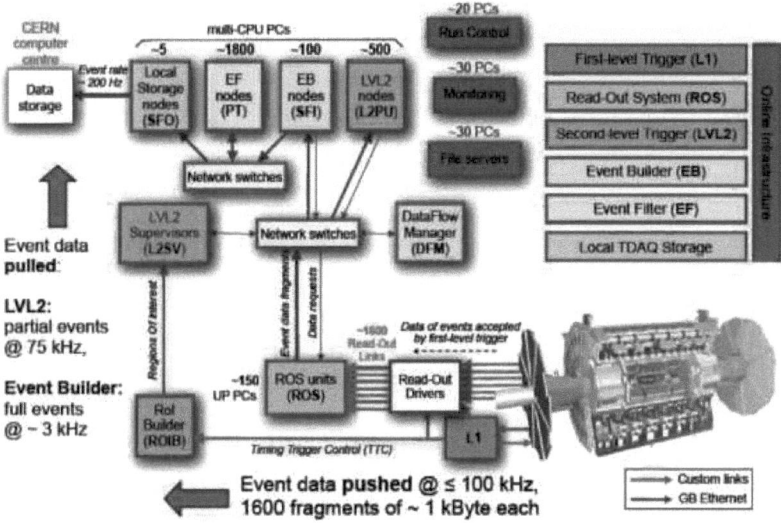

Figure 27: Schematic overview of the ATLAS trigger and DAQ system [62].

5.6.1 Level 1 (L1) Trigger

The level 1 (L1) trigger makes an initial selection based on reduced information from a subset of detectors. High transverse momentum (high p_T) muons are identified using only the trigger chambers, RPCs in the barrel and TGCs in the end-caps. The calorimeter selections are based on reduced granularity information from all the calorimeters (EMC and HC; barrel, end-cap and forward). Objects searched for by the calorimeter trigger are high p_T electrons and photons, jets, and τ leptons decaying into hadrons, as well as large missing and total transverse energies.

In the case of the electron/photon and hadron/τ triggers, energy isolation cuts can be applied. Trigger information is provided for a number of sets of p_T thresholds. The missing and total scalar transverse energies used in the L1 trigger are calculated by summing over trigger towers. In addition, a trigger on the scalar sum of jet transverse energies is also available. The L1 trigger decision is based on combinations of objects required in coincidence or veto. The maximum rate at which the ATLAS front-end systems can accept L1 triggers is limited to 75 kHz [51, p. 25].

L1 Electron/Photon Trigger The L1 electron trigger uses information of the calorimeter, considering a window of 4×4 cluster towers in the EMC and the HC in the pseudo-rapidity range of $|\eta| < 2.5$ [51, p. 360].
The trigger decision is made from four requirements:

- a 2×2-tower EMC cluster, used to identify the position of candidate RoIs;

- a 2×1 or 1×2-tower EMC cluster, used to measure the transverse energy of EM showers; the most energetic of the four possible combinations of the 4×4 window is used;

- a ring of 12 electromagnetic towers surrounding the cluster window, used to test isolation in the EMC Calorimeter;

- the 16 hadronic towers behind the electromagnetic clusters and the isolation ring are used to test the isolation in the hadronic calorimeters.

Since no track information is taken into account at the L1 trigger step, the trigger cannot distinguish between electrons and photons.

L1 Muon Trigger The L1 muon trigger requires hits in three elements of the RPC and TGC of the muon spectrometer. The predicted development of the muon track defined by the first station is extrapolated as a straight line to the second and the third one and has to

5 THE ATLAS DETECTOR 85

match the hit position in the second and the third station around a coincidence window [51, p. 354].

5.6.2 Level 2 (L2) Trigger

The L2 trigger makes use of "region-of-interest" (RoI) information provided by the L1 trigger. This includes information on the spatial position (η and φ) and p_T of candidate objects and energy sums ($\not{E}_T$ vector and scalar E_T value).
The RoI data is sent by L1 to L2, for all events selected by the L1 trigger. The L1 trigger passes only the data that is required in order to make the L2 decision. The L2 trigger has access to all of the event data, if necessary with the full precision and granularity. Usually only a few per cent of the full event data is required for L2 decisions [51, p. 25].

Concerning the energy sum triggers ($\not{E}_T$, total scalar E_T), only a limited improvement is possible using the RoI mechanism. The energy sum values from L1 are provided to L2 and refinements can be made to correct, e.g. for high p_T muons (the L1 $\not{E}_T$ trigger uses only calorimeter information, so muons contribute to the observed $\not{E}_T$) or for saturated trigger tower signals.
The L1 trigger makes RoI information available for all the objects that contributed to the event being selected: these are called primary RoIs. In order to allow for additional requirements to be made at L2, the L1 trigger provides RoI information for objects that did not contribute to the selection of the event. Such RoIs, typically for objects of relatively low p_T, are called secondary RoIs [51, p. 26].

L2 Electron/Photon Trigger At the L2 level, electrons and photons are distinguished using shower shape information in the EMC in the corresponding RoI [51, p. 376]. The electron hypothesis is confirmed if there is an associated track extracted from TRT/pixel detector tracking information. At L2 level, the full calorimeter granularity is taken into account.

L2 Muon Trigger In a first step, the RoI information from the L1 muon trigger is refined by assembling all candidate tracks from L1 to a curved muon track including MDT RoIs in order to be able to determine the momentum of the L2 muon object [51, p. 371].
In a second step, tracks in the MDTs are identified regarding the refined RoIs. A preliminary muon momentum for the L2 trigger decision is reconstructed from MDT, TGC and RPC information.

5.6.3 Event Filter (EF) Trigger

After L2, the last stage of the on-line selection is performed by the EF. It employs off-line algorithms and methods, adapted to the on-line environment, and uses the most up to date calibration and alignment information and the magnetic field map. The EF provides the final selection of physics events which will be written to the mass storage for subsequent full off-line analysis. The output rate from L2 should then be reduced by one order of magnitude, giving $\approx 100\,\text{Hz}$, corresponding to an output data rate of $\approx 100\,\text{MB/s}$ if the full event data is to be recorded [51, p. 26].

5.6.4 Data Storage and Processing

The amount of collision data taken by the ATLAS detector is by far too large to be stored locally at CERN. A worldwide, hierarchical storage organization distributes data to computer facilities located all over the world.
First, the raw event data is stored temporarily at the CERN server farm, called Tier-0, and copied to national Tier-1 facilities. Derived datasets from the raw data are copied to Tier-2 facilities afterwards.

Tier-2 facilities also provide CPU power for data analyses. Institutions can install local Tier-3 computing facilities with a smaller capacity to whom ATLAS has access. The communication between the computers connected to Tier-0...3s is controlled by the LHC computing grid (LCG) environment [63, pp. 1].

The storage size needed per event is gradually decreased by transforming the raw data to more specialized data formats. First, the raw data is converted to *Event Summary Data* (ESD) where the raw data is transformed to reconstructed physical objects. The ESDs are further compressed to *Analysis Object Data* (AOD) which is applicable for specific analyses.

While AODs are still universally utilizable for any analysis, only specific objects contained in the AODs needed for specific analyses are selected and stored in *Derived Physics Data* (DPD) [63, pp. 7].

The DPDs can be distinguished in two classes:

- **D2PD**: The object model in D2PDs is compatible with those in AODs so that analyses do not have to be adapted for the usage of D2PDs instead of AODs.

- **D3PD**: The content of D2PDs is converted to a ROOT tree [64] that makes data access easy to handle.

The event reconstruction is performed by the Athena framework [63, pp. 27] which is also responsible for the event simulation chain (see Sec. 10.1).

Part II
Analysis

6 Monte Carlo Generators

Monte Carlo (MC) generators compute the amplitude of a desired physical process, calculate its cross section and generate simulated events by randomly choosing points in the phase space of incoming and outgoing particles and taking into account the PDFs of the colliding particles (see Sec. 2.2.1). Since this event simulation happens at a relatively large energy scale, the process simulated by the full ME calculation is often referred to as *hard process*.

A various number of different MC generators is used within the ATLAS experiment, most of them generate events at the leading-order (LO) level, a few of them on the next-to-leading order (NLO) level.

6.1 MC@NLO

In order to achieve a higher precision in the calculation of the total cross section and the kinematic distributions of outgoing particles, a next-to-leading order calculation (NLO) is preferred instead of using leading-order (LO). Since there are no next-to-leading-log (NLL) parton showering and hadronization programs available[9], parton matching is difficult. MC@NLO [65] is a program that calculates

- the phase space of several processes whose matrix elements (ME) already exist at the NLO;

- final state partons that directly can be interfaced to LL PS programs
  (HERWIG/PYTHIA);

---
[9] JetPhox is one of the first approaches to NLL.

6 MONTE CARLO GENERATORS

- can handle hard gluon radiation (since divergences cancel out at NLO level).

MC@NLO uses the *subtraction method* and generates simulated events with positive and negative event weights. It is used for simulating top quark pair and single top production, in particular for the production of the nominal $t\bar{t}$ MC sample used in this analysis (see Tab. 29) and for the simulation of single top quark processes (see Tab. 48).

6.2 POWHEG

POWHEG is a NLO MC generator that simulates events at exact NLO level but avoids negative event weights [66]. POWHEG is used to evaluate systematic uncertainties caused by either uncertainties in the NLO calculation or uncertainties due to the usage of different parton showering programs.

Probe $t\bar{t}$ MC samples had been produced externally with POWHEG for analyzing systematic uncertainties related to the NLO MC modeling of top quark pair production (see Sec. 12.3.7) and for the studies of the impact of the choice of different parton showering/hadronization programs (cf. 6.6.1 and 6.6.2).

6.3 ALPGEN

ALPGEN is a LO MC simulation program that provides several full ME calculations with the emission of a selectable number of additional partons [67].

W and Z production with the associated production of jets is simulated with ALPGEN (see Sec. 10.3).

6.4 AcerMC

AcerMC is a LO MC simulation program especially designed to evaluate and simulate typical background processes that occur in hadron colliders [68].
Various $t\bar{t}$ MC samples have been produced externally using AcerMC for systematic studies related to ISR and FSR radiation (see Sec. 12.3.7).

6.5 WHIZARD

WHIZARD is a multi-purpose, LO Monte Carlo event generator with automated matrix element generation [69] which takes into account all possible contributing Feynman diagrams. It supports up to eight particles in the final state and possesses a multi-channel phase-space integration (VAMP algorithm). The matrix element is calculated by the O'Mega ME generator [70]. The simulated $t\bar{t}\gamma$ signal events used in this thesis are generated with WHIZARD.

6.5.1 The Matrix Element Generator O'Mega

With an increasing number of final state particles, the number of contributing Feynman diagrams increases approximately by a double factorial law. Many of these diagrams are redundant and lead to a large loss of numerical precision in the calculation of the matrix element since many representations cancel out analytically but will not do so when integrating over the phase space numerically. Furthermore, many diagrams between external lines having an off-shell propagator can be combined to a *One Particle Off-Shell Wave Function* (1POW). At tree level, the set of all 1POWs for a given set of external momenta can be constructed recursively.

O'Mega considers only combinations of diagrams (*keystones*) with a distinct combination of momenta. $P(n) = 2^n - 1$ distinct momenta can be formed from n external momenta. Exploiting the equation of motions avoids double-counting of keystones [71]. The choice of a certain keystone shaped from a set of Feynman diagrams is not unique, the keystone closest to the center of the overall diagram is used. O'Mega uses Direct Acyclical Graphs (*DAGs*) to avoid redundancies in joining Feynman graphs.

The ME calculation of O'Mega is build of the following steps:

Grow: starting from the external particles, the tower of all 1POWs up to a given height is built (the height is less than the number of external lines for asymmetric keystones and less than half of that for symmetric keystones) and is translated to the equivalent DAG D.

Select: from D, all possible flavored keystones for the process under consideration are determined and the 1POWs appearing in them.

Harvest: a sub-DAG $D^* \subseteq D$ consisting only of nodes that contribute to the 1POWs appearing in the flavored key-centered stones is constructed.

Calculate: the 1POWs as specified are multiplied by the keystones and sum the keystones.

6.5.2 Phase Space Integration

WHIZARD uses the VAMP algorithm for phase space integration which is an advancement of the VEGAS algorithm [72]. In order to calculate the integral $I(f) = \int_M d\mu(p) f(p)$ over a phase space (manifold M) numerically (expectation value $E(f)$), the integrand $f(x)$ is usually weighted by some probability density $g(x)$:

$$E(f) = \left\langle \frac{f}{g} \right\rangle = \frac{1}{N} \sum_{i=1}^{N} \frac{f(p_i)}{g(p_i)}. \qquad (125)$$

$g(p)$ is to be chosen in a way that the variance of $E(f)$

$$V(f) = \frac{1}{N-1}\left(\left\langle\left(\frac{f}{g}\right)^2\right\rangle - \left\langle\frac{f}{g}\right\rangle^2\right) \qquad (126)$$

will become minimal. If f is a strongly fluctuating function, this optimization of g is indispensable for obtaining a useful accuracy. Typical causes for large fluctuations are integrable singularities of f or μ inside of M or non-integrable singularities very close to M.

Manual optimization of g is often too time consuming, in particular if the dependence of the integral on external parameters (in the integrand and in the boundaries) is not known a-priori. Adaptive numerical approaches are suitable to solve this problem. Optimizing g numerically has been solved for factorizable distributions by the Vegas algorithm [73].

6.6 Parton Showering and Hadronization

A ME event generator simulates events for a desired process up to a very limited number of final state particles. These particles include fermions (leptons and quarks) and/or gauge bosons like photons, gluons and massive gauge bosons. In an ideal case, the ME event generator would calculate the matrix elements up to stable hadrons, so up to particles that can reach the detector. Due to the very high complexity of such calculations and the high number of final state particles this is not feasible.

For this reason, parton shower programs like HERWIG [74] or PYTHIA [75] approximately calculate the radiation of gluons and photons from quarks and thus distribute the high energies and large momenta of final state particles from the ME event generator down to an energy scale where quarks hadronize to mesons and baryons ($\Lambda = \Lambda_{\text{QCD}}$).

6 MONTE CARLO GENERATORS

Furthermore, PS programs have to deal with the simulation of initial and final state radiation.

6.6.1 HERWIG

HERWIG is a general-purpose Monte Carlo event generator that includes the simulation of hard lepton-lepton, lepton-hadron and hadron-hadron scattering and soft hadron-hadron collisions [74]. It uses the parton shower approach for initial and final state QCD radiation, including color coherence effects and azimuthal correlations both within and between jets. It matches first-order matrix elements with parton showers and correctly treats spin correlations and heavy quark decays. Final state parton showering (PS) in HERWIG is generated by a coherent branching algorithm were the fractions of energy transmitted to split particles are calculated by DGLAP splitting functions (see Sec. 2.2.1). The emission angles are distributed according to Sudakov form factors, which normalize the branching distributions to give the probabilistic interpretation needed for a MC simulation. A cluster model is used to perform the final hadronization [74, pp. 24]. Finally, mesons and baryons are simulated according to the quark content in the clustered parton showers.

HERWIG is the standard PS program used for all nominal MC samples considered in this thesis.

6.6.2 PYTHIA

PYTHIA contains a variety of matrix elements such as hard and soft QCD processes, W/Z production, prompt photons production etc. Within the ATLAS simulation chain, it is mainly used for the simulation of PS and hadronization.

The parton showering contains of a hard process contains initial and final state radiation. Depending on the incoming and outgoing particles, QCD and QED radiation can occur.

The default model of string fragmentation implemented in PYTHIA is the *Lund model* [76].

6.6.3 PHOTOS

PHOTOS is an algorithm that provides improved prediction of QED interference and multiple-photon radiation by calculating radiative corrections in decays of resonances such as Z boson or W decays [77]. PHOTOS is usually interfaced to HERWIG.

6.6.4 JIMMY

JIMMY is a library of routines that can be linked with HERWIG. JIMMY generates multiple parton scattering events in hadron-hadron, photon-photon or photon-hadron collisions and is used to simulate underlying events (UE) [78].

7 Physical Objects and Reconstruction

During reconstruction, the signals from the individual detector parts have to be assigned to meaningful physical objects and particles like electrons, photons, muons, jets and missing transverse energy ($\displaystyle{\not}E_T$). In this section, the reconstruction of objects and the performance of the reconstruction algorithms[10] and identification are presented.

7.1 Electrons

The electron reconstruction in the ATLAS detector uses EMC cluster information and tracking information from the inner detector [79]. The reconstruction algorithm starts with a preliminary set of seed clusters in the EMC and searches for 3×5 cell windows in the middle layer calorimeter using a sliding window technique. The seed clusters are required to have a threshold energy of $2.5\,\mathrm{GeV}$. Duplicate clusters are removed from nearby seeds.

For $|\eta| < 2.5$, tracks of charged particles can be reconstructed by the ID. In this region, electrons are identified by associating at least one track to a cluster by extrapolating the tracks from their last point of measurement to the second layer in the EMC. The η and ϕ coordinates of the extrapolated track are compared to the corresponding values of a seed cluster in the second layer. If the differences of the coordinates are within a given threshold, the track is considered to be matched to the cluster. The difference in the ϕ coordinate depends on the magnetic field in the ID (bremsstrahlung), leading to an asymmetric $\Delta\phi$ distribution.

If more than one track matches to the electron cluster, the tracks are ordered according to their reconstruction quality. Tracks having hits in the SCT have priority over tracks without SCT hits. Tracks

---
[10] A reconstruction algorithm is often referred to as "author".

without SCT hits are considered to originate more likely form photon conversions.

If a seed cluster is classified as an electron candidate, the cluster is recomputed using a 3 × 7 sliding window in the middle layer barrel EMC (5 × 5 sliding window in the end-cap EMC). The initial 3 × 5 seed cluster has to be explicitly a subset of the new 3 × 7 window. The reconstructed cluster energy is corrected and finally the electron four-momentum is calculated using both the cluster and track information. The energy is taken as a weighted average between cluster energy and track momentum. The η and ϕ coordinates are always taken from the track.

Electrons with a low transverse momentum of a few GeV are reconstructed with a track as a starting point. The track is then used as seed to find a cluster in the EMC.

7.1.1 Electron Identification

Electrons can be separated into three classes of identification: loose, medium and tight. The classification is based on evaluating shower variables derived from the EM shower in the EMC (see Tab. 4). The identification is performed by a multivariate analysis (MVA) in 10 bins of η and 11 bins of E_T (from 5 GeV to 80 GeV).

Loose Selection Loose electrons are identified using the EM shower variables R_{had1}, R_{had}, R_η and $w_{\eta 2}$. The electron candidate has to be detected within the acceptance region of the detector ($|\eta| < 2.47$).

Medium Selection The medium electron requires a preceding loose identification. w_{stot} and E_{ratio} are taken into account for the MVA. The impact parameter has to be $d_0 < 5$ mm. The track matching requires $\Delta \eta_1 < 0.01$. There has to be at least one overall hit in the pixel detector and ≥ 7 hits in the SCT.

Tight Selection A tight electron requires the medium identification. E/p is considered in addition in the MVA. A tighter restriction on $\Delta\eta_1$ is required ($\Delta\eta_1 < 0.005$) and an additional constraint of $\Delta\phi_2 < 0.02$. Furthermore, there is a requirement on n_{TRT} and the ratio of high threshold and overall TRT hits (n'_{TRT}/n_{TRT}). The tight electron selection rejects electrons whose tracks match photon conversions.

7.1.2 Converted Photons

There is an ambiguity between prompt electrons and electrons from photon conversions. Both kinds of objects exhibit a track that can be associated to an EMC cluster. Such electrons from photon conversion are treated as electrons either so that almost all converted photons are also reconstructed as an electron. This ensures a high electron reconstruction efficiency but leads to a large contamination of electrons with converted photons. Further particle identification algorithms are used to distinguish between converted photons and prompt electrons.

7.2 Muons

There are three types of reconstructed muons at the ATLAS detector: *standalone muons, combined muons* and *segment tagged muons* [80, 81]. For each of these types of muons, there exist two different kinds of algorithms, *MuId* and *Staco*. In this analysis, only the MuId algorithm is used.

Standalone Muons The muon trajectory is only reconstructed in the muon spectrometer (MS). The muon momentum measured in the MS is corrected for the parametrized energy loss of the muon in the calorimeter in order to obtain the muon momentum at the interaction point. The direction of flight and the impact parameter of the

Type	Description	Variable name
Hadronic leakage	Ratio of E_T in the first layer of the hadronic calorimeter to E_T of the EMC cluster (used for $0.8 < |\eta| < 1.37$)	R_{had1}
	Ratio of the overall E_T in the hadronic calorimeter to E_T of the EMC cluster (used for $0.8 < |\eta| < 1.37$)	R_{had}
EMC second layer	Ratio of cell energies (3×7 vs. 7×7 window size) in bins of η	R_η
	Lateral width of the shower	$w_{\eta 2}$
EMC first layer	Total shower width	w_{stot}
	Ratio of the difference of energy associated with the largest and second largest energy deposit over the sum of these energies	E_{ratio}
Track quality	Number of hits in the pixel detector	n_{PIX}
	Number of hits in the SCT	n_{SCT}
	Transverse impact parameter	d_0
Track matching	$\Delta\eta$ between the EMC cluster and the track	$\Delta\eta_1$
	$\Delta\phi$ between the EMC cluster and the track	$\Delta\phi_2$
	Ratio of the EMC cluster energy and the track momentum	E/p
TRT	Total number of hits in the TRT	n_{TRT}
	Ratio of the number of high-threshold and the total number of TRT hits	n'_{TRT}/n_{TRT}
b-layer	Number of hits in the b-layer of the pixel detector	$n_{PIX,b}$

Table 4: Definition and description of the EM shower variables used as electron identification criteria described in [79].

muon at the interaction point are determined by extrapolating the spectrometer track back to the beam line.

Combined Muons The momentum of the stand-alone muon is combined with the momentum measured in the inner detector. The muon trajectory in the inner detector also provides information about the impact parameter of the muon trajectory with respect to the primary vertex.

Segment Tagged Muons A trajectory in the inner detector is identified as a muon if the trajectory extrapolated to the muon spectrometer can be associated with straight track segments in the precision muon chambers.

In the MuId algorithm, muon track candidates are obtained by comparing matching track segments in the MS by a track finding procedure using Hough transforms. Track patterns from unbent tracks (x-z plane) are matched to those obtained from the bent (x-y plane) track patterns. A combined track fit to all muon hits in the ID and the MS is performed.

7.3 Jets

In contrast to leptons and photons, quarks do not exist as single particles. Final state quarks of a physical process begin to radiate gluons and, after the particles affected by QCD have undershot an energy less than Λ_{QCD}, begin to hadronize. Typically, all hadrons are located within a cone around the initial quark track, called a jet. The final goal is to draw conclusions from jets on the initial kinetic quark properties.

7.3.1 Infrared and Collinear Safety

Quarks and gluons can radiate additional gluons that are either very soft ($E_g \to 0$) or collinear to the initial particle ($\Delta\theta \to 0$). From the theoretical point of view, the representing Feynman Diagrams of infrared and collinear fragmentation lead to divergences but are part of the physical reality. The jet algorithm has to take into account such processes and should always yield the same jets, regardless of having soft or collinear radiation or not. In this case, the jet algorithm is called *infrared and collinear safe*.

7.3.2 Jet Algorithms

Jet algorithms combine calorimeter towers of particles generated during the QCD fragmentation and hadronization process to physical objects with a pre-defined cone size.

Basically, jet algorithms can be distinguished in to classes: *cone algorithms* and *cluster algorithms* [82]. Cone algorithms consider the topology of a jet, i.e. they are based solely on geometrical assumptions on the jet shape. In a first step, the algorithm groups particles to *protojets*, i.e. groups of nearby particles with their sum of transverse energies above a certain energy threshold. From those protojets, the one with the highest transverse energy is selected as seed. From this seed, a cone with radius R is drawn. Within this cone, the E_T weighted centroid is calculated that results in a new cone with radius R. This procedure is repeated until the cone (centroid) is stable. The whole algorithm is performed until all protojets are part of a jet. Finally, overlapping jets are either split or merged.

For the split and merge procedure, an overlap threshold f is introduced. Then, the highest energetic protojet P_1 is selected and compared to any protojet P_n that share a constituent P' with P_1. If there ratio $E_T(P')/E_T(P_n)$ is below the threshold ratio f, the jet is split along the axis of the center of two protojets, otherwise the two protojets are merged. If there is no overlap, the considered protojet is promoted to be a final jet.

Such cone jet algorithms are neither infrared nor collinear safe. An adaptive algorithm that omits a fixed seed by considering all possible stable cones ("SIS-Cone") is infrared and collinear safe up to a certain theoretical fixed order calculation.

Cluster algorithms are based on grouping together two nearby objects. There are two different kinds of cluster algorithms: k_t algorithms group closest objects first while the anti-k_t jet algorithm groups two objects with the highest p_T[11]. Since cluster algorithms do not assume a fixed cone size, final jets do not need to be merged or split, every object is uniquely assigned to one jet. Cluster jets are infrared and collinear safe.

The k_t algorithm considers the distance d_{ij} between two objects defined as
$d_{ij} = \min(k_{ti}^2, k_{tj}^2) \Delta R_{ij}/R^2$ with $\Delta R_{ij} = (y_i - y_j)^2 + (\phi_i - \phi_j)^2$ and R being the cone size of the jet. If $d_{ij} = k_{ti}^2$, i.e. if the considered minimum is already the protojet itself, the algorithm stops. Otherwise, the protojets i and j are combined and the procedure is repeated.

The anti-k_t algorithm works exactly the same way, the only difference is that the exponents in the definition of d_{ij} have a negative sign: $d_{ij} = \min(k_{ti}^{-2}, k_{tj}^{-2}) \Delta R_{ij}/R^2$ [83].

In scenarios with high pile-up contribution, the cluster algorithms are behaving the most stable way.

The anti-k_t jet algorithm is entirely used for ATLAS top quark analyses.

7.3.3 Jet Reconstruction

Jets for top quark analyses with the ATLAS detector are reconstructed using topoclusters [84]. Topoclusters are three-dimensional objects seeded by a calorimeter cell with $|E_{\text{cell}}| > 4\sigma_{\text{cell}}$ above the noise threshold where σ_{cell} is the RMS of the energy distribution measured for multiple events. The noise depends on the sampling layer of the cell and the position in $|\eta|$. Neighboring cells with $|E_{\text{cell}}| > 2\sigma_{\text{cell}}$ are

---

[11] "k_t" refers to the transverse momentum of a single particle or calorimeter cell considered for jet reconstruction.

then added to the cluster, increasing the size of the cluster until no nearest-neighbor cell is found above the $2\sigma_{\text{cell}}$ threshold anymore. Finally, all nearest-neighbor cells surrounding the cells of the cluster are added to it in order to improve energy resolution. Clusters with negative energy are removed from the jet reconstruction.

7.3.4 Jet Energy Calibration

After jets have been reconstructed, their measured deposited energy has to be mapped back to the original parton (hadronic) energy. For this procedure, several prescriptions exist. The final jet energy after applying the calibration from these prescriptions is called the *jet energy scale* (JES).

First, jets are reconstructed at the electromagnetic scale [85], which is the basic signal scale for the ATLAS calorimeters. It accounts correctly for the energy deposited in the calorimeter by electromagnetic showers. The energy scale of the electromagnetic calorimeters is calibrated using the invariant mass of $Z \rightarrow ee$ events from collision data. The hadronic jet energy scale is restored using data-derived corrections and calibration constants derived from the comparison of the reconstructed jet kinematics to those of the corresponding truth level jet in MC studies.

The jet calibration corrects for detector effects that affect the jet energy measurement [85]:

1. partial measurement of the energy deposited by hadrons (calorimeter non-compensation),

2. energy losses in inactive regions of the detector (dead material),

3. energy deposits from particles not contained in the calorimeter (leakage),

7 PHYSICAL OBJECTS AND RECONSTRUCTION

4. energy deposits of particles inside the truth jet that are not included in the reconstructed jet,

5. signal losses in calorimeter clustering and jet reconstruction.

Top quark physics analyses with the ATLAS experiment use the *EM+JES calibration scheme* which applies jet-by-jet corrections as a function of the jet energy and pseudo-rapidity to jets reconstructed at the electromagnetic scale. The additional energy due to multiple proton-proton interactions within the same bunch crossings (pile-up) is corrected before the hadronic energy scale is restored, so that the derivation of the jet energy scale does not depend on the number of additional interactions measured anymore.

The EM+JES calibration scheme consists of three subsequent steps:

1. the average additional energy due to pile-up is subtracted from the energy measured in the calorimeters using correction constants extracted from an in-situ measurement,

2. the position of the jet is corrected such that the jet direction points to the primary vertex of the interaction instead of the geometrical center of ATLAS detector,

3. the jet energy and position as reconstructed in the calorimeters are corrected using constants derived from the comparison of the kinematics of reconstructed jets and corresponding truth jets in MC simulations.

In addition, the pseudo-rapidity of jets calibrated on the EM+JES scale is corrected, since the jet direction is biased towards better instrumented regions of the calorimeter [85, p. 5].

7.3.5 Jet b-Tagging

The top quark decays nearly solely into a W boson and a b quark (Sec. 2.4). Hence, the detection of one or more jets formed by the hadronization of a b quark is a strong indicator for a typical top quark decay signature.

b quarks form B mesons that have a relatively long lifetime; the B^0 meson has a mean lifetime of $\tau_{B^0} \approx 1.5\,\text{ps}$ [86]. Assuming speed of light, this corresponds to an average distance of $d = 450\,\text{µm}$ in the rest frame of the meson. This translates to an effective transverse distance of $\langle l \rangle \approx 3\,\text{mm}$ ($\langle l \rangle = \beta\gamma c\tau$ with $\gamma = E_{B^0}/m_{B^0}$ and $\beta = \sqrt{1 - 1/\gamma^2}$) for a jet transverse momentum of $p_T = 50\,\text{GeV}$, leading to a displaced decay vertex [87, p. 398].

There are two basic approaches of evaluating the displaced vertex: The first method calculates the *impact parameter* of the secondary vertex. The transverse impact parameter d_0 is the distance of closest approach of the track to the primary vertex in the r-ϕ plane. The longitudinal impact parameter z_0 is the z coordinate of the track at the point of closest approach in the r-ϕ plane. The tracks from B-hadron decay products tend to have rather large impact parameters which can be distinguished from tracks stemming from the primary vertex [87, p. 398].

There are three algorithms based on the impact parameter [87, p. 407]: IP1D relies on the longitudinal impact parameter, IP2D on the transverse impact parameter and IP3D uses two-dimensional histograms of the longitudinal versus transverse impact parameters, taking advantage of their correlations.

The second method (SV-tagger) reconstructs the displaced vertices explicitly. Two-track pairs from tracks far away from the primary vertex are formed. The tracks from the two-track vertices are combined into a single inclusive vertex by an iterative χ^2 optimization [87, p. 407]. The SV1 tagger considers a 2D distribution of two discriminating variables, SV2 uses a 3D histogram taking into account three discriminating variables.

Furthermore, there is the "JetFitter" algorithm, which exploits the topological structure of weak b- and c-hadron decays inside the jet. It assumes that the b- and c-hadron decay vertices lie on the same line defined by the b-hadron flight path. All particle tracks stemming from either the b- or c-hadron decay thus intersect the b-hadron flight axis [88, 89].

A combined, advanced b-tagging algorithm combines the IP3D and the JetFitter algorithm by a neural network analysis (`JetFitterCOMBNN`) [90]. This advanced tagger is used for the analysis presented in this thesis.

b-**Tagging Calibration** The performance of b-tagging depends on two parameters: The b-tag efficiency and the probability to falsely identify a jet as a b-tagged jet (*mistag rate*). The b-tag efficiency has been evaluated using both the $p_{T,\text{rel}}$ method which exploits the harder p_T spectrum of muons that stem from B meson decays and the usage of top quark pair reconstruction where two b-jets are expected to appear in the event signature [91].

7.4 Photons

If during electron reconstruction there is no track associated to the EMC seed cluster or a track associated that points to a conversion vertex, the cluster is considered to be a photon candidate (*unconverted photon*) [92–94].

An additional noise cleaning where noisy calorimeter cells that appear on top of those already been marked as noisy cells in a database, is applied in order to avoid photons reconstructed due to such additional cells when performing the sliding window search algorithm.

Clusters of unconverted photons that lack a track are recomputed by applying energy calibrations and corrections due to longitudinal and transverse energy losses (*second sampling*).

Converted photon candidates have either one (*single track conversions*) or two (*double track conversions*) tracks associated to their cluster [93]:

Double conversions are reconstructed by performing a constrained vertex fit using the information of the two tracks under the assumption that the photon is a massless particle.

The reconstructed track in single track conversions is usually located at larger radial positions inside the tracker. Here, a vertex fit cannot be performed, the conversion vertex is then placed at the location of the first measurement of the participating track, typically when one of the two tracks failed to be reconstructed either because it is either soft (*asymmetric conversions* where one of the two tracks has $p_T < 0.5\,\text{GeV}$), or when the two tracks are very close to each other (*symmetric conversions*: the two tracks have similar transverse momenta) and cannot be separated adequately.

Single and double track conversion candidates are classified as photons, if they can be matched to a reconstructed EM calorimeter cluster. A track is considered to be matched to an electromagnetic cluster if its impact point after extrapolation from its last measurement to the calorimeter second sampling is within a certain range in η-ϕ space from the cluster center.

Almost all converted photons will end up inside the reconstructed electron collection and have therefore to be re-identified as photons (see Sec. 7.4.1).

7.4.1 Photon Recovery

After the first step of electron/photon discrimination described above, most of the converted photon candidates are treated as electron candidates. Furthermore, there are a few unconverted photon candidates that have erroneously matched tracks from other sources than an elec-

7 PHYSICAL OBJECTS AND RECONSTRUCTION 107

tron/conversion track. These photon candidates are identified by a *photon recovery* procedure.

For electron candidates with conversion vertices, the track matching best is compared with the track(s) originating from the best conversion vertex candidate matched to the same cluster. If the track matching best coincides with a track coming from the conversion vertex, then this electron is treated as a converted photon, except in the case of a double-track conversion vertex candidate where the coinciding track has a b-layer hit, while the other track lacks one. If the track does not coincide with any of the tracks assigned to the conversion vertex candidate, then it is kept as an electron, unless the track p_T is smaller than that of the converted photon candidate.

Converted photons can also be recovered from electron candidates if there is no conversion vertex assigned: if the originally reconstructed electron has a best matched track that is made of only TRT hits (TRT-only track) with $p_T > 2\,\text{GeV}$ and $E/p < 10$ (E being the cluster energy, p the track momentum) it is considered to be a converted photon, regardless of whether a conversion vertex candidate has been matched to its EM cluster or not. In the second case a vertex is assigned to the first hit of the track resulting in a single-track converted photon candidate.

The cluster of the converted photon is rebuilt with a larger window size of 3×7 (5×5) in the barrel (end-cap) EMC [92].

All electrons matched to a track reconstructed in the TRT only and with $p_T < 2\,\text{GeV}$ are automatically considered to be unconverted photon candidates. In addition, electrons that failed to be considered as converted photon candidates, and for which their best matched track has $p_T < 2\,\text{GeV}$ or $E/p > 10$, will also be treated as unconverted photon candidates.

7.4.2 Photon Identification

The photon identification algorithms use rectangular cuts on calorimetric shower variables in order to distinguish between prompt photons and photon fakes from photons that have their origin in jet fragmentation. Photons are classified as "loose" and "tight" photons [95]. The calorimetric variables used in the photon selections can be grouped in three main categories: hadronic leakage, variables using the second longitudinal compartment (middle layer) of the EMC and variables using the first longitudinal compartment (strip layer) of the EMC. A complete list of these shower variables and their impact on the photon identification is given in Tab. 5.

The tight photon requirements are optimized to reject π^0 mesons misidentified as photons. The fine granularity of the EMC strip layer provides a good $\gamma \leftrightarrow \pi^0$ separation.

The tight cuts are separately optimized for unconverted and converted photon candidates, since the electron tracks of converted photons are bent in the ϕ plane due to the magnetic field of the solenoid. EM showers from converted photons are therefore more spread than those of unconverted photons.

The optimization has been performed using a MVA technique, providing optimized tight photon identification in seven bins of $|\eta|$ for converted and unconverted photons separately. An identification efficiency of $\approx 85\%$ is achieved. Fig. 28 to 31 show the dependence of photon identification efficiencies on E_T in four bins of $|\eta|$ for converted and unconverted photons separately.

At the time this thesis was created, no estimation of the photon efficiency from data had been available.

7 PHYSICAL OBJECTS AND RECONSTRUCTION

Category	Description	Name	Loose	Tight
Acceptance	$|\eta| < 2.37$, $1.37 < |\eta| < 1.52$ excluded		✓	✓
Hadronic leakage	Ratio of E_T in the first sampling of the hadronic calorimeter to E_T of the EM cluster (used over the range $|\eta| < 0.8$ and $|\eta| > 1.37$)	R_{had_1}	✓	✓
	Ratio of E_T in all the hadronic calorimeter to E_T of the EM cluster (used over the range $|\eta| > 0.8$ and $|\eta| < 1.37$)	R_{had}	✓	✓
EM middle layer	Ratio in η of cell energies in 3×7 versus 7×7 cells	R_η	✓	✓
	Lateral width of the shower	w_2	✓	✓
	Ratio in ϕ of cell energies in 3×3 and 3×7 cells	R_ϕ		✓
EM strip layer	Shower width for three strips around maximum strip	$w_{s,3}$		✓
	Total lateral shower width	$w_{s,tot}$		✓
	Fraction of energy outside core of three central strips but within seven strips	F_{side}		✓
	Difference between the energy of the strip with the second largest energy deposit and the energy of the strip with the smallest energy deposit between the two leading strips	ΔE		✓
	Ratio of the energy difference associated with the largest and second largest energy deposits over the sum of these energies	E_{ratio}		✓

Table 5: Description of photon shower variables used for photon identification [95].

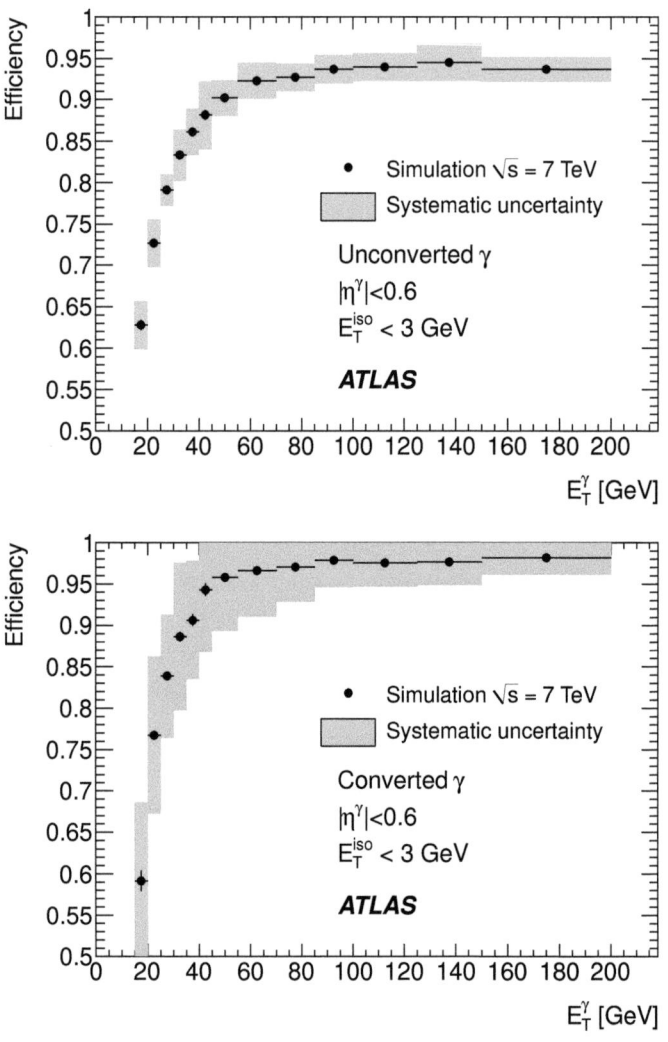

Figure 28: Identification efficiency of unconverted and converted photons in bins of E_T derived for the isolated di-photon cross section measurement [96]. The pseudo-rapidity range of $|\eta| < 0.6$ is shown. The yellow areas indicate the efficiency uncertainties.

7 PHYSICAL OBJECTS AND RECONSTRUCTION 111

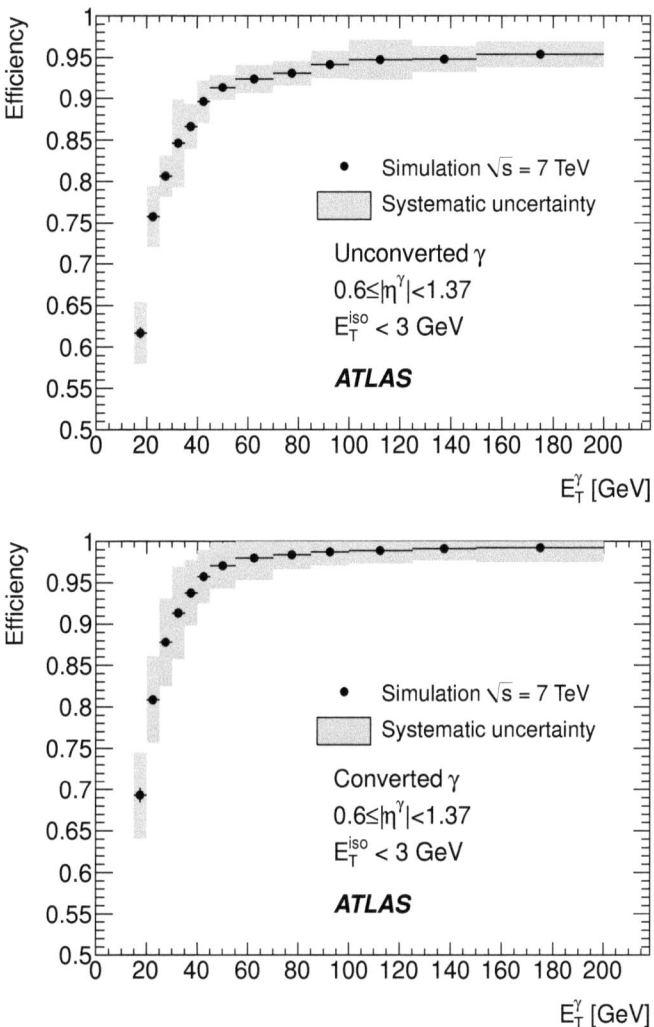

Figure 29: Identification efficiency of unconverted and converted photons in bins of E_T derived for the isolated di-photon cross section measurement [96]. The pseudo-rapidity range of $0.6 \leq |\eta| < 1.37$ is shown. The yellow areas indicate the efficiency uncertainties.

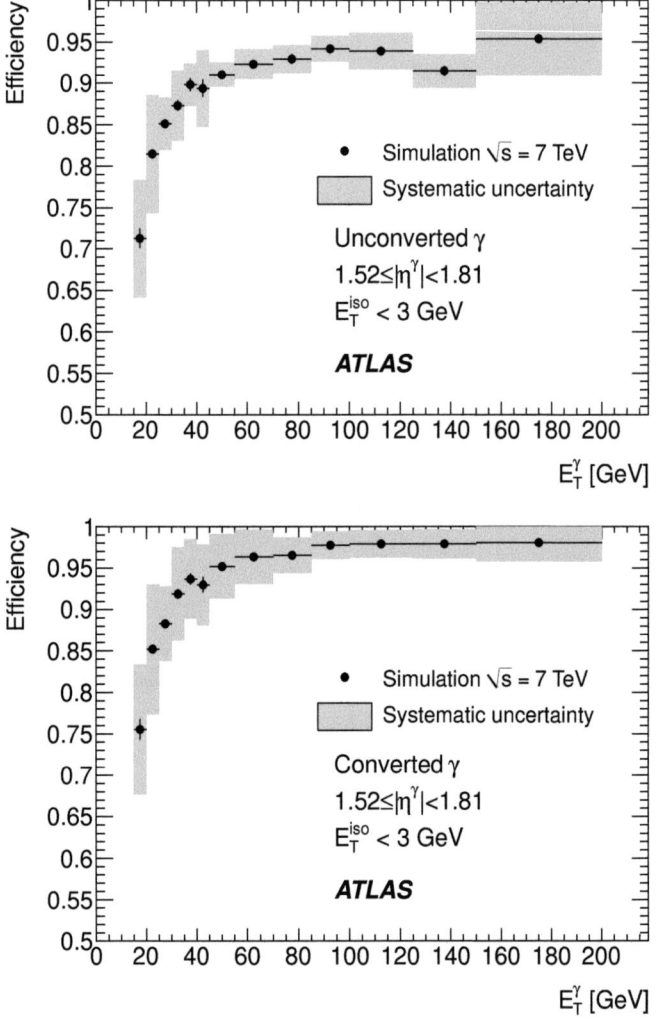

Figure 30: Identification efficiency of unconverted and converted photons in bins of E_T derived for the isolated di-photon cross section measurement [96]. The pseudorapidity range of $1.52 \leq |\eta| < 1.81$ is shown. The yellow areas indicate the efficiency uncertainties.

7 PHYSICAL OBJECTS AND RECONSTRUCTION

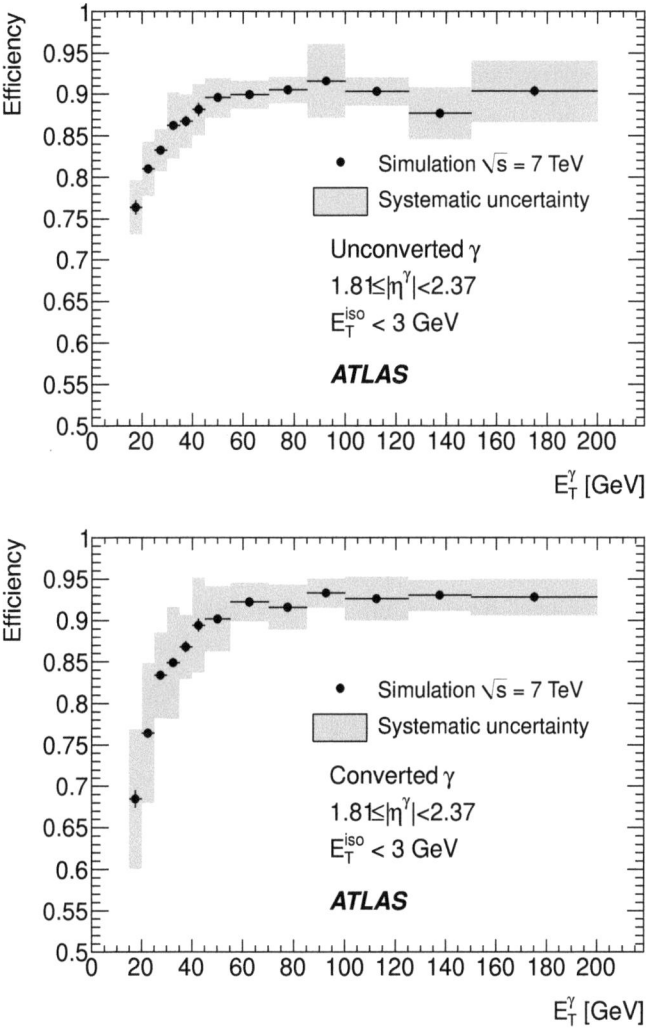

Figure 31: Identification efficiency of unconverted and converted photons in bins of E_T derived for the isolated di-photon cross section measurement [96]. The pseudorapidity range of $1.81 \leq |\eta| < 2.37$ is shown. The yellow areas indicate the efficiency uncertainties.

7.5 Missing Transverse Energy

The $\not{E}_T$ reconstruction used in ATLAS physics analyses includes contributions from transverse energy deposits in the calorimeters and muons reconstructed in the muon spectrometer [97]. The two $\not{E}_T$ components are calculated as:

$$\not{E}_{x(y)} = \not{E}_{x(y)}^{\text{calo}} + \not{E}_{x(y)}^{\mu} \, . \tag{127}$$

The calorimeter term is defined as:

$$\not{E}_x^{\text{calo}} = -\sum_{i=1}^{N_{\text{cell}}} E_i \sin\theta_i \cos\phi_i \, , \quad \not{E}_y^{\text{calo}} = -\sum_{i=1}^{N_{\text{cell}}} E_i \sin\theta_i \sin\phi_i \, , \tag{128}$$

where E_i, θ_i and ϕ_i are the energy, the polar angle and the azimuthal angle of cells over the pseudo-rapidity range $|\eta| < 4.5$. Noise contributions are reduced by using only cells belonging to three-dimensional topoclusters. These topoclusters are seeded by cells with a deposited energy of $|E_i| > 4\sigma_{\text{noise}}$, and are built by iteratively adding neighboring cells with $|E_i| > 2\sigma_{\text{noise}}$ and finally adding all neighbors of the accumulated cells (cf. Sec. 7.3.3).

Only cells that belong to reconstructed physical objects are included in the calculation of $\not{E}_T^{\text{calo}}$. In order to avoid double counting of calorimeter cells, objects are taken into account by the following order: electrons $\rightarrow$ photons $\rightarrow$ hadronically decaying τ-leptons $\rightarrow$ jets $\rightarrow$ muons. Cells belonging to topoclusters not associated with any such objects are also taken into account in the $\not{E}_T$ calculation (*cell-out terms*). With the given order of physical objects, $\not{E}_T$ is calculated by:

$$\not{E}_{x(y)} = \not{E}_{x(y)}^{e} + \not{E}_{x(y)}^{\gamma} + \not{E}_{x(y)}^{\tau} + \not{E}_{x(y)}^{\text{jets}} + \not{E}_{x(y)}^{\text{soft jets}} + \not{E}_{x(y)}^{\text{calo }\mu} + \not{E}_{x(y)}^{\text{cell out}} \, . \tag{129}$$

Contributions from jets are calculated from jets with $p_T > 20\,\text{GeV}$, those from soft jets from jets with $7\,\text{GeV} < p_T < 20\,\text{GeV}$.

7 PHYSICAL OBJECTS AND RECONSTRUCTION

The MET term from muons is obtained from the negative sum of energy contributions of muons in the calorimeter,

$$\not{E}^\mu_{x(y)} = - \sum_{\text{muons}} p^\mu_{x(y)} \qquad (130)$$

taking into account only the pseudo-rapidity range of $|\eta| < 2.5$. Remaining adjacent cells that are not included in the objects are collected in the $\not{E}^{\text{cell out}}_{x(y)}$ term.

7.5.1 $\not{E}_\text{T}$ Calibration and Calibration Schemes

The missing transverse energy depends on the choice of a *calibration scheme* which accounts for dead material in the detector and detector response. The available schemes are as follows:

Global calibration (GCW) Cell weights are calculated from a global cell energy density.

Local calibration LCW Weights are computed from properties of individual clusters.

The $\not{E}_\text{T}$ definition in Eq. (129) depends on the definition of the physical objects involved and will depend on whether "loose", "medium" or "tight" object definitions have been chosen on the one hand and on the calibration scheme on the other hand. The standard `MET_RefFinal` $\not{E}_\text{T}$ definition uses medium electrons with $p_\text{T} > 10\,\text{GeV}$, tight photons with $p_\text{T} > 10\,\text{GeV}$ and tight τ leptons with $p_\text{T} > 10\,\text{GeV}$ (LCW calibration).
The jet terms are calculated using the anti-k_t algorithm with a cone size of $R = 0.6$ with JES correction applied and are included with the LCW calibration.

The $E_{x(y)}^{\text{cell out}}$ term uses the LCW scheme. The contribution of these cell energies is corrected using the p_T information from tracks that do not reach the calorimeter.

In the LCW (GCW) based $\not{E}_T$ configuration, all objects contribute with the LCW (GCW) calibration. There is an additional EM based configuration, where all terms are calculated on the EM scale. The latter one is used for the analysis presented in this thesis.

7.6 Isolation

The isolation requirement on objects that are reconstructed in a narrow cone, like photons, electrons and muons, is an important indicator to distinguish the prompt generation of these particles in the hard process from those produced as decay products from hadrons in jets.

Electron isolation can be determined as additional energy detected in a cone with a given cone size around the electron cluster ($E_{T,\text{iso}}$) or by summing all tracks within a cone around the electron cluster ($p_{T,\text{iso}}$) [79].

The calorimetric isolation discriminator $E_{T,\text{iso}}^{R_0}$ is computed from the reconstructed transverse energy in a cone of half opening angle R_0 around the axis of the electron candidate. The energy of the electron itself is excluded. While a larger cone will contain more energy in case of mis-identified jets, a smaller cone is more robust against energy depositions from pile-up events.

The track based isolation $p_{T,\text{iso}}^{R_0}$ is the scalar sum of transverse momenta of tracks in a cone of size R_0 around the electron. In contrast to the calorimetric isolation, neutral particles do not contribute to this quantity. The advantage is that the track quality criteria can be applied in order to reject tracks from pile-up vertices. Only tracks with $p_T > 1\,\text{GeV}$ are considered, which have a hit in the innermost pixel detector layer, at least 7 hits in silicon detectors and a transverse impact parameter less than 1 mm.

7 PHYSICAL OBJECTS AND RECONSTRUCTION 117

The isolation variables for photons are obtained the same way as for electrons [94]. In contrast to electrons, no constraint on the longitudinal impact parameter is applied per default[12]. Tracks within $\Delta R < 0.1$ to the photon are also required to be not matched to a conversion vertex.

The isolation criteria for muons follow those of electrons either, using the ID tracking and calorimetry information of the combined muons [80].

7.6.1 Calorimeter Isolation Corrections

As quoted, the usage of the calorimeter isolation is not as robust against pile-up effects as the track isolation. Furthermore, a lateral leakage of cell energy that belongs to the electron into the surrounding isolation cone might occur. Additional noise bursts in the EMC can increase the measured isolation. These effects have been corrected after the standard electron reconstruction procedure [98].

---

[12]In the framework of this analysis, this definition was changed and the track isolation was recalculated.

8 Dataset

The dataset analyzed in this thesis corresponds to an integrated luminosity of $1035.4\,\text{pb}^{-1}$, recorded by the ATLAS detector from LHC pp collisions at a CMS energy of $\sqrt{s} = 7\,\text{TeV}$, collected from March 2011 until June 2011. During this time, the instantaneous luminosity was steadily increased. Fig. 32 shows the progress of the amount of collected data vs. time in 2011 and the daily instantaneous luminosity vs. time, Fig. 33 depicts the distribution of the average number of bunch crossings $\langle n_{\text{BX}} \rangle$.

An increasing instantaneous luminosity may lead to pile-up effects which emerge in additional locations of particle interactions. Hence, additional primary vertices can be reconstructed.

Each LHC fill contains one or more ATLAS *runs*, each ATLAS run contains several *luminosity blocks* (LB). Runs with similar collider and/or detector conditions are grouped in *periods*. The data used for this analysis contains data from period B2 until H4.

During data taking, the proton bunch settings were changed as well as the detector conditions. Furthermore, the focusing of the proton beams at the IP was improved and the bunch spacing was reduced.

A detailed overview on the runs contained in the data periods and the progress of increasing the number of bunches and bunch trains is given in Tab. 6.

Although approximately $5\,\text{fb}^{-1}$ of data were recorded in 2011, only $1.04\,\text{fb}^{-1}$ have been taken into account for the analysis presented in this thesis. This limitation on the amount of considered data has two reasons:

1. When the analysis software framework used for this thesis was programmed, the ATLAS Offline Software release 16 (`rel16`) was the current version. For `rel16`, only $2.05\,\text{fb}^{-1}$ of 2011 data had been processed.

2. Relevant input parameters from external analyses like scale factors have been evaluated for `rel16` only for $1.04\,\text{fb}^{-1}$.

8.1 Data Streams

Within the ATLAS top quarks physics working group, reconstructed and recorded events have been classified in three categories according to the targetted decay channel, i.e. whether to use the lepton+jets (semi-leptonic), di-lepton or the hadronic decay mode. The events that fulfil the requirements of one of these three categories are exported into a corresponding class of files (*data stream*) [99].

e/γ Stream The e/γ stream requires at least one reconstructed electron with a transverse momentum of $p_\text{T} > 20\,\text{GeV}$ or one electron with $p_\text{T} > 13\,\text{GeV}$ and an additional requirement with two alternatives, either requiring the missing transverse energy to be $\not{E}_\text{T} > 20\,\text{GeV}$ or requiring two arbitrary leptons (electron or muons, reconstructed with the MuId algorithm) with each having a transverse momentum of $p_\text{T} > 13\,\text{GeV}$. This stream is used for e+jets and di-lepton analyses. Since the e/γ stream considers electrons reconstructed from both the track based and the calorimeter based algorithm (see Sec. 7.1), events where a photon meets the electron requirements are included in this stream either.

μ Stream The definition for the μ stream is very similar to that of the e/γ stream and requires one muon with a transverse momentum of $p_\text{T} > 18\,\text{GeV}$ or one muon with $p_\text{T} > 13\,\text{GeV}$ and an additional requirement with two choices, either requiring the missing transverse energy to be $\not{E}_\text{T} > 20\,\text{GeV}$ or requiring two arbitrary leptons (electron or muons), each of them having a transverse momentum of $p_\text{T} > 13\,\text{GeV}$. This stream is used for μ+jets and di-lepton analyses.

Jet/$\not{E}_T$ Stream The jet/$\not{E}_T$ stream requires at least four jets with a transverse momentum of $p_T > 20\,\text{GeV}$ and two jets with $p_T > 40\,\text{GeV}$ among them or, alternatively, at least five jets with $p_T > 20\,\text{GeV}$ (and no requirement on a number of jets with a higher p_T). This stream is used for analyses considering the hadronic decay mode and particularly in this analysis to extract the isolation spectrum of photons originating from jet fragmentation (see Sec. 11.3.3).

8.2 Good Run Lists

During data taking, not all detector parts of ATLAS might be operating or several modules of one detector could have been switched off. Events recorded during such periods cannot be reconstructed at an adequate quality and have to be removed. For this purpose, the luminosity blocks in a run a flagged if they meet the requirements on the event quality. This flagging is stored in *good run lists* (GRLs).

The GRLs considered in this analysis reduce the bare integrated luminosity delivered to ATLAS from $L = 1.291\,\text{fb}^{-1}$ to $L_{\text{eff}} = 1.035\,\text{fb}^{-1}$, hence only $\approx 80\,\%$ of all recorded events could be used for the analysis.

8 DATASET

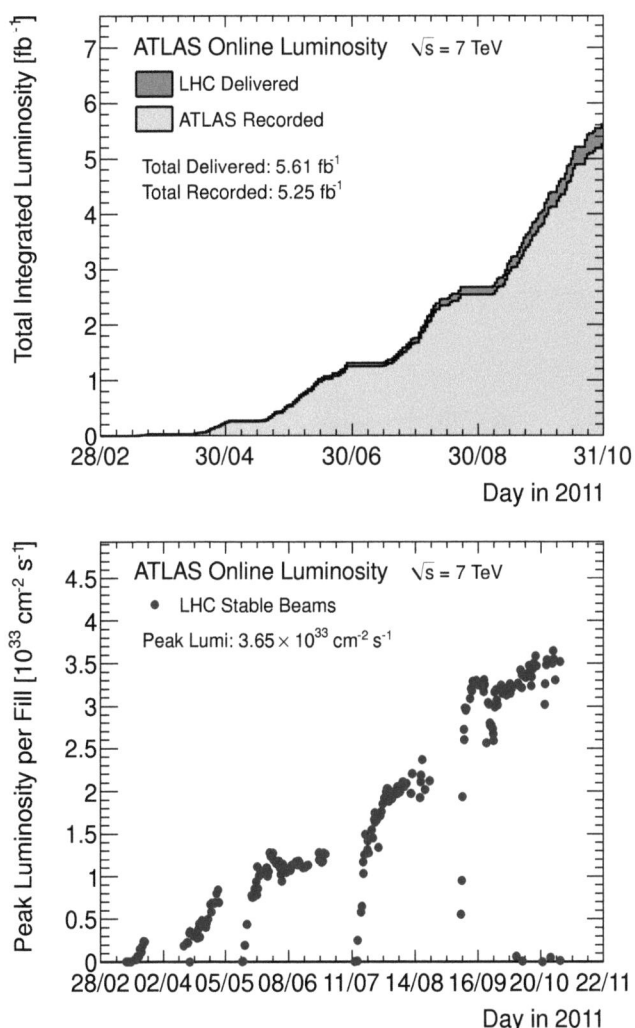

Figure 32: *Upper pane:* Integrated luminosity of data collected in 2011 (left plot) and instantaneous luminosity of the LHC within that time (right plot) [100].

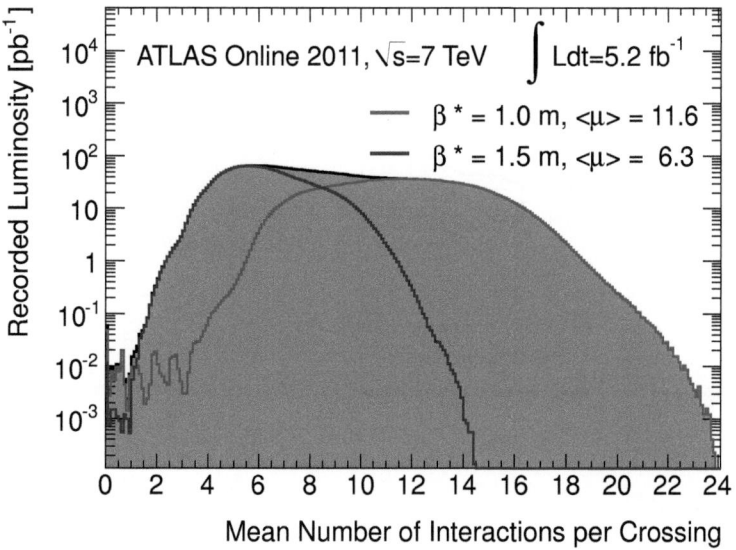

Figure 33: Distribution of the average number of bunch crossings $\langle n_{\mathrm{BX}} \rangle$ for all data collected by the ATLAS detector in 2011 (5.2 fb^{-1}) [100].

8 DATASET

period	run numbers	bunches	bunch trains	bunch spacing [ns]	$\int L\,dt$ [pb^{-1}]
B2	178044, 178047, 178109	194	9	75	13.33
D1	179739, 179725, 179710	214	8	50	12.60
D2	179804, 179771	322	11	50	11.45
D3	180144, 180139, 180124, 180122, 179940, 179939, 179938	322	11	50	39.95
D4	180149, 180153, 180212	424	14	50	38.36
D5	180225, 180241, 180242	424	10	50	32.19
D6	180309, 180400, 180448	598	11	50	32.34
D7	180481	598	11	50	24.56
E1	180614, 180636	598	11	50	32.80
	180664, 180710, 180776	700	14	50	20.72
F1	182013	14	1	–	0.03
	182032, 182034	214	8	50	0.51
F2	182161	424	10	50	3.44
	182284, 182346, 182372	700	10	50	39.49
	182424, 182449, 182450, 182454, 182455, 182456, 182486	874	11	50	98.29
F3	182516, 182518, 182519	874	11	50	20.42
G1	182726	874	11	50	6.65
G2	182747, 182766	874	11	50	36.23
	182787, 182796, 182879, 182886	1042	13	50	82.18
G3	182997, 183003, 183021	1042	13	50	79.86
G4	183038, 183045, 183078, 183079, 183081, 183127, 183129, 183130	1042	13	50	122.16
G5	183216, 183272, 183286, 183347	1042	13	50	107.56
G6	183391, 183407, 183412, 183426, 183462	1042	13	50	145.33
H1	183544, 183580, 183581, 183602	1042	13	50	54.93
H2	183780	1041	12	50	48.36
H3	183963, 184022, 184066, 184072	1180	13	50	68.72
H4	184074, 184088, 184130	1180	13	50	71.70
	184169	1318	13	50	47.22
				$\sum =$	1291.38

Table 6: Description of the data periods used in this analysis [101, 102].

9 Object Definitions and Event Selection

The reconstructed objects described in Sec. 7 are of general purpose for any analysis and provided ATLAS-wide. Each analysis has specific requirements on these objects in order to optimize the purity of an event selection, like the setting of thresholds on the transverse momentum, suitable triggers, the minimum number of selected objects or a minimum/maximum $\not{E}_T$.

Furthermore, there are detector effects that might occur after the reconstruction algorithms were applied on raw detector data like problems with hardware components.

The object definitions and event selection for the $t\bar{t}\gamma$ cross section measurement are defined in this section. They are based on the common object definitions and the single lepton + jets event selection of the ATLAS top quark working group [103] with slight modifications.

9.1 Electrons

Electrons are required to have a transverse momentum of $p_T > 25\,\text{GeV}$. The transverse momentum is calculated from the electron cluster energy and the direction of the electron track only:

$$p_T = \frac{E_{cl}}{\cosh \eta_{tr}}. \tag{131}$$

The electron cluster energy is corrected according to Sec. 9.5.

Electrons are required to be identified as tight electrons (see Sec. 7.1.1) and have to be reconstructed by either the track based algorithm or the track based and cluster algorithm simultaneously (`author == 1 || author == 3`). Electrons that coincide with photon objects after the recovery procedure (see Sec. 7.4.1) are removed.

The pseudo-rapidity range of the EM cluster is restricted to be $|\eta_{cl}| < 2.47$ and the transition region between barrel and end-cap EMC is

9 OBJECT DEFINITIONS AND EVENT SELECTION

excluded ($1.37 < |\eta_{\text{cl}}| < 1.52$). The cluster isolation (see Sec. 7.6) within a cone of $R < 0.2$ is required to be $E_{\text{T,iso}}^{\text{cone20}} < 3.5\,\text{GeV}$. The isolation is corrected for pile-up effects, cone leakage and calorimeter noise (see Sec. 7.6.1).

Electrons with parts of its cluster lying in EMC regions with not properly working optical transmitters (OTX) are removed.

Additionally, for MC simulation, electrons that are reconstructed in the region where 6 FEBs of the LAr calorimeter failed on April 30th in 2011 are removed (see Sec. 9.8).

9.2 Muons

Combined muons reconstructed with the MuId algorithm with a transverse momentum of $p_{\text{T}} > 20\,\text{GeV}$ are selected and have to pass the "tight" identification criteria. The pseudo-rapidity must be in the central part of the detector ($|\eta| < 2.5$).

Track and calorimeter isolation criteria for a cone of $R = 0.3$ are applied ($p_{\text{T,iso}}^{\text{cone30}} < 4\,\text{GeV}$ and $E_{\text{T,iso}}^{\text{cone30}} < 4\,\text{GeV}$).

Additional requirements on the muon ID tracks have to be fulfilled: At least one hit in the b-layer of the pixel detector is required unless the track transverses a region, where the b-layer is not instrumented or not working properly. The overall sum of hits in the pixel detector and the number of crossed pixel sensors not working properly has to be greater than one.

Similarly, the overall sum of hits in the TRT and the number of crossed TRT modules not working properly must be ≥ 6, the number of holes in the pixel detector and the TRT must be < 2.

Furthermore, the expected number of hits in the TRT that contribute to the reconstruction of the muon track is optimized for the central and the outer regions of the TRT. If n_{TRThits} is the number of TRT hits on the muon track, $n_{\text{TRToutliers}}$ the number of TRT outliers on the muon track, and $n = n_{\text{TRThits}} + n_{\text{TRToutliers}}$:

- For muons with $|\eta| < 1.9$, $n > 5$ and $n_\text{TRToutliers} < 0.9 \cdot n$ is required,

- for muons with $|\eta| \geq 1.9$, $n_\text{TRToutliers} < 0.9 \cdot n$ is only required if $n > 5$.

Muons that overlap with jets within a cone of $\Delta R < 0.4$ are removed in order to avoid muons from hadron decays if the jet has a transverse momentum of $p_\text{T} > 20\,\text{GeV}$ and positive energy ($E > 0.0\,\text{GeV}$). Due to a bug in the reconstruction software for MC simulation, a correct trigger matching cannot be applied for muons with $p_\text{T} > 150\,\text{GeV}$. Hence muons with a transverse momentum larger than $150\,\text{GeV}$ are excluded from the analysis.

9.3 Triggers and Trigger Matching

The single lepton + jets analysis requires one triggered lepton with a certain p_T threshold. This threshold has to be below the actual p_T requirement in the event selection in order to ensure that all selected leptons have been identified by the trigger and the event selection is not strongly biased by the trigger efficiency.

High level offline EF triggers are chosen for the e+jets and the μ+jets channel.

For the e+jets channel, a single electron trigger with a threshold of $p_\text{T} > 20\,\text{GeV}$ (EF_e20_medium) is required where the trigger object has to pass the medium criteria. This trigger requires the L1 trigger L1_EM14 and the succeeding L2 trigger L2_e20_medium with the same p_T threshold as the final EF trigger [101, 102].

For the μ+jets channel, a single muon trigger with a threshold of $p_\text{T} > 18\,\text{GeV}$ is used (EF_mu18). This EF trigger requires the L1 trigger L1_MU10 and the succeeding L2 trigger L2_mu18.

9 OBJECT DEFINITIONS AND EVENT SELECTION

The selected electron has to match the according trigger object both in data and MC simulations, muons are required to match the according trigger object in MC only[13] [103].

For the electron trigger matching, the distance ΔR in $\eta - \phi$ space between the electron and all electron trigger objects that have passed the EF_e20_medium criteria is calculated. If any of the trigger objects fulfils $\Delta R < 0.15$, the electron is tagged to match the trigger.

Muons from MC simulations are treated with a similar procedure. Additionally, the matching procedure is performed also on the L2 trigger objects. Both the EF and L2 trigger matching have to fulfil the $\Delta R < 0.15$ criterion. On top of the trigger matching, the EF muon tracks are tested for a high-p_T trigger hypothesis since the track qualities depend on their kinematic quantities. The trigger hypothesis is defined as follows:

- For muon trigger objects with $|\eta| < 1.05$, the transverse momentum has to be $p_T > 17.53\,\text{GeV}$.

- For muon trigger objects with $1.05 \leq |\eta| < 1.5$, the transverse momentum has to be $p_T > 17.39\,\text{GeV}$.

- For muon trigger objects with $1.5 \leq |\eta| < 2.0$, the transverse momentum has to be $p_T > 17.34\,\text{GeV}$.

- For muon trigger objects with $|\eta| \geq 2.0$, the transverse momentum has to be $p_T > 17.28\,\text{GeV}$.

9.4 Photons

Photons are required to be identified as tight photons and to exhibit a transverse momentum of $p_T > 15\,\text{GeV}$, where p_T is recalculated

---
[13]This recommendation was established due to a software bug

from the cluster energy and cluster position obtained from the second sampling:

$$p_T(\gamma) = \frac{E_{cl,S2}}{\cosh \eta_{S2}}. \tag{132}$$

The pseudo-rapidity of the cluster from the second sampling has to be $|\eta_{S2}| < 2.37$ and the transition region between barrel and end-cap EMC is excluded ($1.37 < |\eta_{S2}| < 1.52$).

The photon has to be reconstructed either as unconverted or converted photon (author == 0x4 || author == 0x10) and has to pass the photon recovery procedure (Sec. 7.4.1). Photons with parts of its cluster located in EMC regions with optical transmitters (OTX) or high voltage supplies not working properly are removed.

9.4.1 Photon Isolation

The photon track isolation with a cone size of $R = 0.2$ ($p_{T,iso}^{cone20}$) is used as discriminating variable in the analysis presented in this thesis (Sec. 11).

The tracks that are considered for the scalar sum of transverse track momenta (see Sec. 7.6) have to fulfil several quality criteria [104] which are the same for electron and photon isolation:

- The track momentum has to fulfil a minimum threshold of $p_T > 1$ GeV in order to reject low energy tracks from pileup [105].

- The transversal impact parameter is required to be $|d_0| < 1$ mm.

- The sum of hits and holes[14] in the pixel detector and in the SCT has to be $n_{hits} + n_{holes} \geq 7$.

- There has to be a hit and/or a hole in the innermost layer of the pixel detector (*b*-layer) ($n_{hits} + n_{holes} \geq 1$).

---
[14]A hole in the inner silicon detectors are pixels (pixel detector) or strips (SCT) that are expected to contribute a signal to a reconstructed track but do not give a response due to material interactions, semi-conductor inefficiencies or problems with the pattern recognition [106].

Recalculation of Photon Isolation As mentioned in Sec. 7.6, the difference between photon and electron isolation is the additional requirement of the longitudinal impact parameter to be $|z_0| < 1\,\text{mm}$ for electrons. Since the distribution of the prompt photon isolation will later be determined from electrons (see Sec. 11.3.2), the photon isolation has to be recalculated by revisiting all tracks surrounding the photon and discard those with $|z_0| \geq 1\,\text{mm}$ [107].

The algorithm first selects all tracks in the event that fulfil the common track quality criteria described above and are within a cone of $\Delta R < 0.4$ around the photon axis. All possible permutations containing ν^{cone20} tracks out of the selected tracks are created, where ν^{cone20} is the number of tracks contributing to the standard track isolation. The permutation whose sum of track p_T is closest to $p_{\text{T,iso}}^{\text{cone20}}$ (a maximum difference $\Delta p_{\text{T,iso}}^{\text{cone20}} < 100\,\text{MeV}$ is allowed) is considered to be the correct track configuration. From this permutation, the tracks with $|z_0| \geq 1\,\text{mm}$ are removed and the new track isolation is calculated from the remaining tracks.

9.4.2 MC Correction of Photon Shower Variables

The photon shower variables relevant for photon identification, in particular those affecting the profile of the shower shape in the first layer of the EMC, such as R_η, F_side and w_{η_2}, have been found to deviate comparing data and MC simulation, most probably due to an imperfect MC modeling of the lateral distribution of the photon EM shower [108].

Fig. 34 shows this discrepancy between data and MC simulation for the shower shape variables R_η and w_{η_2}. Obviously, a constant shift quantified by the means $\langle \text{DV} \rangle$ of the distributions of data and MC simulation is capable to transform one distribution into the other:

$$\Delta \mu_\text{DV} = \langle \text{DV}_\text{Data} \rangle - \langle \text{DV}_\text{MC} \rangle. \tag{133}$$

These *fudge factors* have been derived for all relevant shower variables and are applied on MC simulation in order to correct the photon ID.

9.5 Energy Rescaling

Due to dead material and intermediate instrumentation in the EMC, the energy deposit of photons and electrons cannot be detected to the full extent. Hence, the reconstructed energy has to be scaled to the real energy based on assumptions on how the energy is deposited and detected in the EMC. A verification of these assumptions and a gradual approach to physical reality can be achieved not before some amount of collision data has been analyzed. In order not to reprocess all recorded data again, a correction to the preliminary estimation of the EMC energy scale has to be applied afterwards.

For this purpose, energy scale factors (SFs) were derived by the ATLAS combined performance groups in order to correct the measured energy to the the real initial energy, once by comparing invariant di-electron masses to the J/ψ and Z boson mass and additionally by investigating the E/p ratio of electrons [109, 110]. The SFs are provided in 58 bins of η, ranging from -4.9 to 4.9.

$Z \to ee$ and $J/\psi \to ee$ decays from collision data were chosen since the invariant masses of both particles are well-known from measurements of other experiments such as LEP. The energy SFs for electrons are extrapolated to photons by comparing both kind of objects in MC simulation samples since there is no event signature where a pure photons sample could be obtained from data.

Systematic uncertainties of this method are caused by MC signal and background modeling, pile-up effects, imperfect knowledge of additional (dead) material, the energy scale of the pre-sampling EMC and effects of the EMC electronics (cross-talk, calibration, read-out non-linearities) and imperfect detector conditions (not all EMC components were working equally reliable).

9 OBJECT DEFINITIONS AND EVENT SELECTION

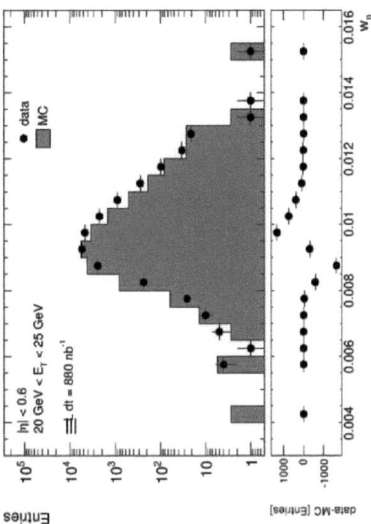

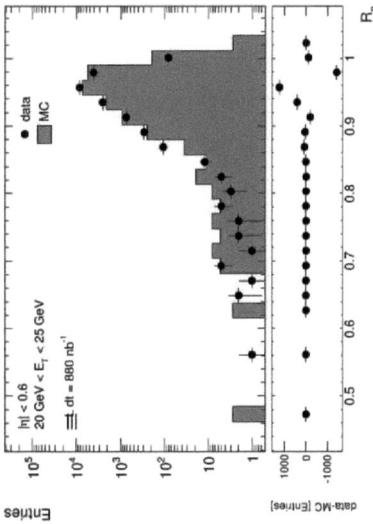

Figure 34: Discrepancies of the photon shower shape variables R_η and w_{η_2} between data and MC simulation [108]. The pull plots in the lower panes imply a constant shift in order to make the distributions match to each other.

The E/p method constraints the energy of electrons to their momenta measured by the tracker system since $E/p \simeq 1$ due to the very small electron mass. Systematic uncertainties of this method stem mainly from the applied fit model, the effect of additional (dead) material and the MC modeling of the Z and/or J/ψ background; systematic uncertainties on distorted ID alignment, finite momentum resolution and bremsstrahlung corrections can affect the momentum measurement. The energy corrections vary from the order of 1 % in the barrel EMC up to 5 % in the end-cap calorimeter. The predicted electron and photon energies by MC simulations are corrected to the improved estimation of the detector resolution also. The energies were smeared using a Gaussian distribution with a width of

$$\sigma = \sqrt{[S(1+\sigma_S)]^2 E_{\text{cl}} + [C(1+\sigma_C)E_{\text{cl}}]^2 - [S \cdot E_{\text{cl}}]^2 - (C \cdot E_{\text{cl}})^2} \tag{134}$$

around the nominal cluster energy [110, 111]. S is a sampling term, C a constant and σ_C, σ_S their respective uncertainties.

9.6 Jets

Jets are reconstructed with the anti-k_{t} algorithm with a cone size of $R = 0.4$ (see Sec. 7.3.2). A transverse momentum of $p_{\text{T}} > 25\,\text{GeV}$ is required, the jet pseudo-rapidity on the EM scale has to be $|\eta_{\text{em}}| < 2.5$, the EM+JES scale η correction is added to the jet η_{em} since this correction is not included for jets calibrated on the EM scale. Jets with negative energy are not considered.

If any jet with $p_{\text{T}} > 20\,\text{GeV}$ and $E > 0\,\text{GeV}$ does not fulfil jet quality cuts (*bad jets*), the whole event is discarded. A jet is defined as a *bad jet* if not all of the following quality cuts are fulfilled [112, 113]:

Removal of HEC spikes The fraction of the energy deposit of the jet in the HEC f_{HEC} exceeds 0.5 and the HEC quality factor Q_{HEC}

9 OBJECT DEFINITIONS AND EVENT SELECTION

is < 0.5. Q_{HEC} is the fraction of energy corresponding to LAr cells with a cell Q-factor greater than 4000. The cell Q-factor measures the difference between the measured pulse shape (a_i^{meas}) and the predicted pulse shape (a_i^{pred}) that is used to reconstruct the cell energy. It is computed as $\sum_{\text{samples}} (a_i^{\text{meas}} - a_i^{\text{pred}})^2$ [114]. Alternatively, the negative energy E_{neg} of a jet can be $E_{\text{neg}} > 60\,\text{GeV}$ (without considering f_{HEC} or Q_{HEC}).

EM Coherent Noise The fraction of jet energy in the EMC f_{EM} is larger than 0.95 and the average jet quality computed as the energy squared Q_{LAr} is larger than 0.8, if the jet $|\eta|$ is < 2.8.

Cosmics and Non-Collision Background The jet time t, computed as the energy squared cells mean time (*timing*), is $|t| > 10\,\text{ns}$. Alternatively, f_{EM} can be < 0.05 and the fraction of charged tracks in a jet f_{ch} is < 0.05. This condition is valid if $|\eta| < 2.0$. Otherwise, if $|\eta| \geq 2.0$, f_{EM} has to be < 0.05. As fourth alternative, the maximum energy fraction in one calorimeter layer F_{max} has to be > 0.99 for $|\eta| < 2.0$.

9.6.1 b-Tagging

b-tagged jets are chosen from the standard jet selection (Sec. 9.6) with the additional requirement of a tag weight $w_{\text{tag}} \geq 0.35$ using the combined advanced `JetFitterCOMBNN` tagger algorithm based on a neural network analysis (see Sec. 7.3.5).
The chosen working point of $w_{\text{tag}} \geq 0.35$ provides a b-tag efficiency of $\approx 70\,\%$ at a purity of $> 91\,\%$ regarding MC@NLO $t\bar{t}$ MC simulation. Contributions from c-quark jets are rejected by a factor of 5, jets from light quarks by a factor of 99 and jets from hadronically decaying τ leptons by a factor of 15 [115].

9.6.2 Electron-Jet Overlap Removal

Since the jet reconstruction algorithm works completely independent from the electron reconstruction algorithm, it cannot distinguish between energy depositions from hadronic jets or from electrons in the calorimeter. Hence, also electrons will be reconstructed as jets.

Since top quarks are reconstructed from jets that have a hadronic origin and to avoid an ambiguity between the electron and the jet collection, jets that have been reconstructed from electrons have to be removed from the selection (*overlap removal*).

For the electron-jet overlap removal, the jet closest in $\eta - \phi$ space to a selected electron is removed, if the spatial distance is $\Delta R < 0.2$. This procedure ensures to keep hadronic jets that are reconstructed nearby the electron jet.

9.6.3 Photon-Jet Overlap Removal

Similarly to electrons, also photons are reconstructed as jets and stored in the jet collection. These photon jets have to be removed from the jet selection.

Fig. 35 shows the ratio of the photon transverse momentum and the jet transverse momentum as a function of the distance between photon and jets in $\eta - \phi$ space. Jets and photons are required to fulfil the object selection criteria described in Sec. 9.6 and Sec. 9.4. The overlap between photons and jets has been studied on the WHIZARD $t\bar{t}\gamma$ MC sample (see Sec. 10.2).

Two effects can be observed in Fig. 35: First of all, most of the entries in the 2D distribution are concentrated in a region with $\Delta R < 0.01$. In this range, the ratio of photon and jet p_T is rather constant. Secondly, the p_T ratio is ≈ 1 in that narrow ΔR region. Above that value, the average p_T ratio decreases linearly until the jet cone size $R = 0.4$ is reached. For $\Delta R > 0.4$, the p_T ratios and ΔR values become completely uncorrelated.

This indicates that, if the jet axis coincides with the photon axis, the jet is in fact a single photon. For increasing ΔR above a threshold value of $\Delta R \gtrsim 0.1$, the photon is carrying a fraction of the total energy of a real hadronic jet.

As consequence, jets that overlap with selected photons within a cone of $\Delta R < 0.1$ in η-ϕ space are removed since these objects are not considered to be real hadronic jets. The overlap removal is performed using the kinematic jet quantities reconstructed on the EM scale.

For remaining photons within a cone of $\Delta R < 0.5$ around the jet axis, the reconstructed jet energy is still distorted by the photon. Besides, the photon isolation is strongly affected by charged tracks and energy depositions in the calorimeter that clearly belong to the jet. Hence, the complete event is discarded if still a selected photon is found within a distance of $\Delta R < 0.5$ to a selected jet.

9.7 Missing Transverse Energy

The missing transverse energy calculated on the EM scale is used, including the photon term (Sec. 7.5). The $\slashed{E}_T$ is recalculated using the scaled electron and photon energies (see Sec. 7.3.4).

9.8 Treatment of EMC Hardware Failures

On April 30th in 2011, six front-end boards of the EMC read-out electronics failed after a LHC machine stop. These six boards could not be brought back to operation and had to be replaced.

As consequence, some regions of the LAr detector were not measuring any energy deposits anymore, a.k.a. the *LAr hole*. Meanwhile, $\approx 400\,\text{pb}^{-1}$ of data were recorded by the ATLAS experiment in this condition.

In order not to loose the data of that period, prescriptions were developed on how to treat the effects of the hardware failure [116]. While

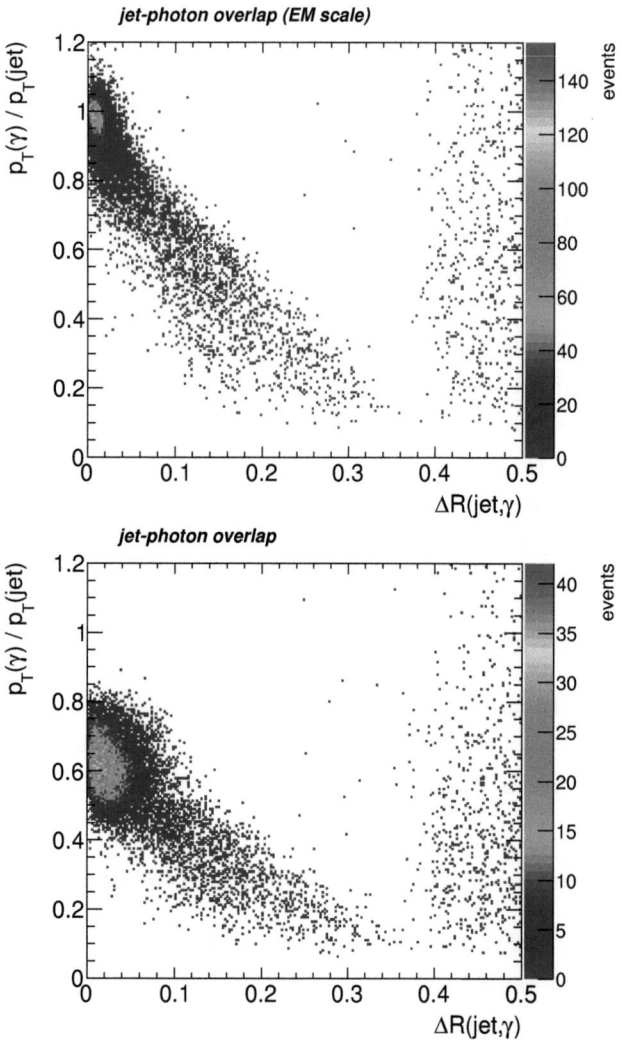

Figure 35: Ratio of photon transverse energy and jet transverse momentum at the EM scale (left) and with the kinematic variables calibrated on the EM+JES scale (right) as a function of the distance between jet and photon in $\eta - \phi$ space. Obviously, the jet is in fact the considered photon for small values of ΔR. The plots have been generated from the $t\bar{t}\gamma$ MC sample.

the affected data was already corrected during recording and reconstruction, the hardware failure has to be emulated in MC simulations.

Treatment of Jets Jets from MC simulations are tested if their transverse momenta exceed a certain threshold. Given $\epsilon_{\text{BCH\_CORR\_CELL}}$ the fraction of broken cells within the jet cone and $\epsilon_{\text{BCH\_CORR\_JET}}$ the predicted fraction of energy of broken cells contributing to the jet [117], the threshold $p_{\text{T}}^{\text{th.}}$ is calculated by

$$p_{\text{T}}^{\text{th.}} = 20\,\text{GeV} \cdot \frac{1 - \epsilon_{\text{BCH\_CORR\_JET}}}{1 - \epsilon_{\text{BCH\_CORR\_CELL}}}. \quad (135)$$

If any reconstructed jet fulfils $p_{\text{T}} > p_{\text{T}}^{\text{th.}}$ and is reconstructed close to the affected detector region, the whole event is discarded.

Treatment of Electrons and Photons In data, electrons and photons that were radiated into the problematic regions could not even be detected. This loss of electrons and photons is emulated in MC simulations by randomly choosing events according to the relative frequency of collected data within the time the hardware problem was present. For these chosen MC events, electrons and photons are flagged to be non-existent if they are reconstructed in the broken detector region [118].

9.9 Event Selection

The $t\bar{t}\gamma$ event selection is chosen in a way to select events with exactly one electron or muon (*single lepton selection*), large missing transverse energy, at least four jets, at least one b-tagged jet and a photon in the final state. Depending on whether an electron or a muon is required in the event selection, two variations of the event selection are considered: the one considering electrons will be referred to as e+jets channel,

the one requiring a muon as μ+jets channel. The object definitions appearing in the event selection have been described above.
The event selection is split into a sequential order of selection steps:

1. The event may not be affected by the detector region with broken FEBs (see 9.8).

2. The event has to be triggered by EF_e20_medium (e+jets channel) or EF_mu18 (μ+jets channel) (see Sec. 9.3).

3. At least one primary vertex with a minimum of five tracks associated to it is required in order to reduce impacts of non-collision background.

4. In the e+jets (μ+jets) channel, at least one selected electron (muon) is required.

5. In the e+jets (μ+jets) channel, exactly one selected electron (muon) is required.

6. In the e+jets (μ+jets) channel, no selected muon (electron) is allowed.

7. The lepton has to match the corresponding trigger objects (see Sec. 9.3).

8. Events with an electron and muon sharing a track reconstructed by the ID are removed. This requirement is tested on muons before the muon-jet ΔR isolation criterion (see Sec. 9.2).

9. Events exhibiting at least one bad jet (Sec. 9.6) with a transverse momentum of at least 20 GeV at the EM scale are removed.

10. In the e+jets (μ+jets) channel, the missing transverse energy is required to be $\not{E}_T > 35$ GeV ($\not{E}_T > 20$ GeV).

9 OBJECT DEFINITIONS AND EVENT SELECTION 139

11. In the e+jets channel, the transverse mass of the leptonically decaying W boson[15] is required to be larger than 25 GeV.

    In the μ+jets channel, the sum of $\not{E}_T$ and the transverse mass of the leptonically decaying W boson must be larger then 60 GeV (triangular cut).

12. At least 2 selected jets are required.

13. At least 3 selected jets are required.

14. At least 4 selected jets are required.

15. At least one of the selected jets has to be tagged as a b-jet (see Sec. 9.6.1).

16. Events with a large noise in the LAr calorimeter are discarded.

17. At least one selected photon is required.

18. In case of the e+jets channel, the invariant mass of the selected electron and selected photon has to be outside the Z boson mass window (86 GeV $< M_{\text{inv}} <$ 96 GeV).

19. No selected photon may be be inside a cone of $\Delta R < 0.5$ around a selected jet, otherwise the event is discarded (The reason is that both the reconstructed jet energy and the photon isolation might be biased in this constellation, see Sec. 9.6.3).

Up to selection step 16, the event selection follows the common event selection for single lepton $t\bar{t}$ analyses, so the $t\bar{t}\gamma$ event selection is in principle a subset of the standard $t\bar{t}$ selection. Hence, the event selection up to that step will be referred to as *preselection*.

The event yields and selection efficiencies for each step of the event selection applied on the $t\bar{t}\gamma$ signal MC sample (see Sec. 10.2), scaled

---

[15]The transverse mass is defined as $M_T = \sqrt{2(p_T(\ell)\not{E}_T - p_x(\ell)\not{E}_x - p_y(\ell)\not{E}_y)}$.

to the expectation for an integrated luminosity of $1.04\,\text{fb}^{-1}$, are shown in Tab. 7.

		e+jets channel		μ+jets channel	
step	brief description	events	efficiency	events	efficiency
0	no cut	2026 ± 3	100 %	2026 ± 3	100 %
1	LAr hardware failure	1798 ± 3	88.7 %	1798 ± 3	88.7 %
2	trigger	555.3 ± 1.5	27.4 %	530.2 ± 1.5	26.2 %
3	prim. vertex with ≥ 5 tracks	553.8 ± 1.5	27.3 %	528.9 ± 1.5	26.1 %
4	≥ 1 lepton	332.5 ± 1.2	16.4 %	336.2 ± 1.2	16.6 %
5	exactly one lepton	321.6 ± 1.2	15.9 %	320.2 ± 1.2	15.8 %
6	no other lepton	295.8 ± 1.1	14.6 %	299.5 ± 1.1	14.8 %
7	trigger matching	295.6 ± 1.1	14.6 %	293.0 ± 1.1	14.5 %
8	e-μ overlap removal	295.6 ± 1.1	14.6 %	293.0 ± 1.1	14.5 %
9	bad jets	295.6 ± 1.1	14.6 %	293.0 ± 1.1	14.5 %
10	$\slashed{E}_T$ requirement	225.9 ± 1.0	11.1 %	271.0 ± 1.1	13.4 %
11	$\slashed{E}_T + M_T(W) / M_T(W)$	195.5 ± 0.9	9.6 %	255.9 ± 1.0	12.6 %
12	at least two jets	186.6 ± 0.9	9.2 %	245.8 ± 1.0	12.1 %
13	at least three jets	155.5 ± 0.8	7.7 %	206.3 ± 0.9	10.2 %
14	at least four jets	97.8 ± 0.6	4.8 %	131.4 ± 0.7	6.5 %
15	≥ 1 b-tagged jet	86.0 ± 0.6	4.2 %	115.0 ± 0.7	5.7 %
16	LAr noise	86.0 ± 0.6	4.2 %	115.0 ± 0.7	5.7 %
17	≥ 1 photon	25.1 ± 0.3	1.24 %	31.3 ± 0.4	1.54 %
18	86 GeV > $M_{\text{inv}}(\gamma, e)$ > 96 GeV	23.7 ± 0.3	1.17 %	31.3 ± 0.4	1.54 %
19	$\Delta R(\gamma, \text{jet}) > 0.5$	20.1 ± 0.3	0.99 %	26.5 ± 0.3	1.31 %

Table 7: Event yields of the $t\bar{t}\gamma$ event selection and corresponding selection efficiencies, shown for the e+jets and the μ+jets channel separately. The event yields are scaled to $1.04\,\text{fb}^{-1}$. Only the statistical uncertainties are shown.

10 Signal and Background Modeling

In order to test the performance of the event selection, the selection applied on data has to be compared with the selection applied on MC simulation. Furthermore, this comparison provides a hint on whether all relevant background processes have been considered and are modeled with an appropriate precision concerning absolute cross section and kinematic spectra.

On the other hand, several aspects of physical reality cannot be modeled well so far, especially QCD related topics like ISR/FSR or multijet production. Besides, jet fragmentation relies on an approximation using Sudakov form factors and the chosen fragmentation model (Sec. 6.6) and the simulation of the hadronization process is difficult. Secondly, the description of the detector relies on many assumptions, like the distribution of additional material in front of the calorimeters and geometric alignment which are parameters that are not known to the full extent [110]. For such reasons, contributions from background processes are estimated using data-driven approaches as often as possible.

This section describes the MC modeling of the $t\bar{t}\gamma$ signal and all relevant background processes and provides a comparison to data for important kinematic variables.

All MC samples are centrally provided by the ATLAS MC Production Group [119]. A single pile-up configuration corresponding to the LHC running with 50 ns bunch spacing is used. For the pile-up simulation, PYTHIA minimum bias events are used and variable pile-up rates are assumed.

The samples were produced with an integrated luminosity of $\sim 10\ldots 15\,\text{fb}^{-1}$ for the signal and background processes. Information about the samples used in this thesis like the cross sections, k-factors and number of events are listed in App. B in detail.

10.1 Simulation Chain

The simulation of MC events for ATLAS analyses starts with generating the four-momenta of final state particles calculated by a MC generator using the exact matrix element for the specific process.
These raw events are passed to a PS program like HERWIG or PYTHIA (see Sec. 6.6) in order to create a set of physical particles like mesons, baryons, leptons and photons which approximately represents all particles as they would reach the detector. As a next step, the detector response is simulated using the Geant4 simulation toolkit [120], taking into account the assumed interaction of the particles with the detector material. The detector response is translated into the corresponding signals the read-out electronics would produce. This step is called *digitization*. As last step, the digitized, simulated data is passed to the standard reconstruction algorithms. A flowchart of the simulation chain is depicted in Fig. 36.

In contrast to data, MC simulations contain both reconstructed objects and particle information of the MC generator and the PS program. This additional MC information is frequently used and referred to as *truth information*.

10.2 Simulation of Signal Events

The $t\bar{t}\gamma$ signal sample has been generated with the WHIZARD MC generator (Sec. 6.5) which allows for the exact calculation of the matrix element of a seven particle final state ($2 \to 7$ process, cf. Sec. 3.2).

A set of non-full-hadronic events has been simulated, i.e. all semi-leptonic and di-leptonic decay modes of top quark pair production with the additional emission of a photon have been considered:

$$pp \to \ell\nu_\ell q\bar{q}'b\bar{b}\gamma \quad \text{and} \quad pp \to \ell\nu_\ell\tilde{\ell}\nu_{\tilde{\ell}}b\bar{b}\gamma \quad \text{with } \ell/\tilde{\ell} = e,\, \mu,\, \tau\,.$$

10 SIGNAL AND BACKGROUND MODELING 143

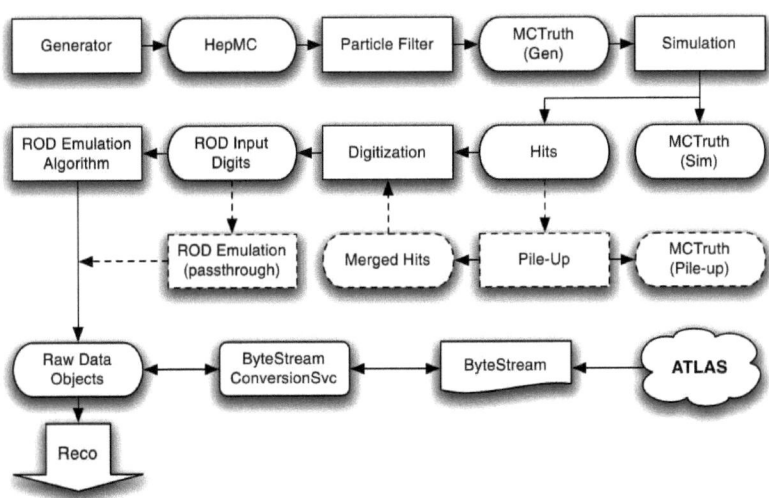

Figure 36: Flowchart of the ATLAS event simulation chain [63, p. 54].

These events have been showered with HERWIG, additional photon radiation from leptons has been calculated by PHOTOS. The CTEQ6L1 PDF set has been applied during event generation. Approximately 500,000 events have been generated and passed to the full ATLAS simulation chain (cf. Tab. 28).

Light quarks (u, d, s, c) and electrons ($e^{\pm}$) are assumed to be massless; the τ lepton has been assigned a mass of $m_\tau = 1.78\,\text{GeV}$. b quarks have a mass of $m_b = 4.2\,\text{GeV}$, the top quark mass is set to $m_t = 172.5\,\text{GeV}$. The mass of muons ($\mu^{+/-}$) is defined as $m_\mu = 105.6\,\text{MeV}$. The mass of the Higgs boson is set to $m_H = 120\,\text{GeV}$[16].

In order to avoid collinear and infrared divergences, the transverse momentum of the photon must be larger than 8 GeV and the invariant masses of the following pairs of particles are required to be larger than 5 GeV:

---

[16] When the $t\bar{t}\gamma$ MC samples were produced, there has not been any evidence for a Higgs boson yet.

- $m_{\text{inv}}(q_1, q_2)$: both quarks from the hadronic W decay,

- $m_{\text{inv}}(q_i, \gamma)$: each quark from the hadronic W decay and the photon,

- $m_{\text{inv}}(l, \gamma)$: the charged lepton from the leptonic W decay and the photon,

- $m_{\text{inv}}(Q_i, \gamma)$: both incoming quarks (except $b/\bar{b}$) and the photon,

- $m_{\text{inv}}(g_i, q_j)$: both incoming gluons and both quarks from the hadronic W decay,

- $m_{\text{inv}}(Q_i, q_j)$: each incoming quark and its corresponding antiparticle from the hadronic W decay, if the anti-particle exists,

- $m_{\text{inv}}(\ell_1, \ell_2)$: the two leptons of di-leptonic events.

The absolute cross section times branching ratio for the generated sample has been calculated as $\sigma_{t\bar{t}\gamma}^{\text{MC}} \times \text{BR} = 0.84\,\text{pb}$. NLO calculations for the process $t\bar{t}\gamma$ for the LHC exist for $\sqrt{s} = 14\,\text{TeV}$ [121]. For a center-of-mass energy of 7 TeV, a k-factor of 2.55 has been estimated for an energy scale of $\mu_R = 2m_t$ and a k-factor of 2.11 for as scenario with $\mu_R = m_t$ respectively [122]. Since the k-factor strongly depends on the phase space definitions, which are not perfectly matching w.r.t. to the theoretical calculations[17], the average $k = 2.33 \pm 0.22$ of both values has been assumed to be the nominal k-factor for the $t\bar{t}\gamma$ normalization.

Fig. 37 shows the distribution of the transverse momenta of truth photons as generated by WHIZARD. The red shaded area below 15 GeV is excluded from the acceptance region, which corresponds to a loss of $\approx 38\,\%$ of all generated events. The measurement of the cross section is extrapolated to the full photon p_T spectrum simulated by WHIZARD.

---

[17]The calculation of k-factors for the WHIZARD $t\bar{t}\gamma$ phase space cuts is currently in progress.

10 SIGNAL AND BACKGROUND MODELING

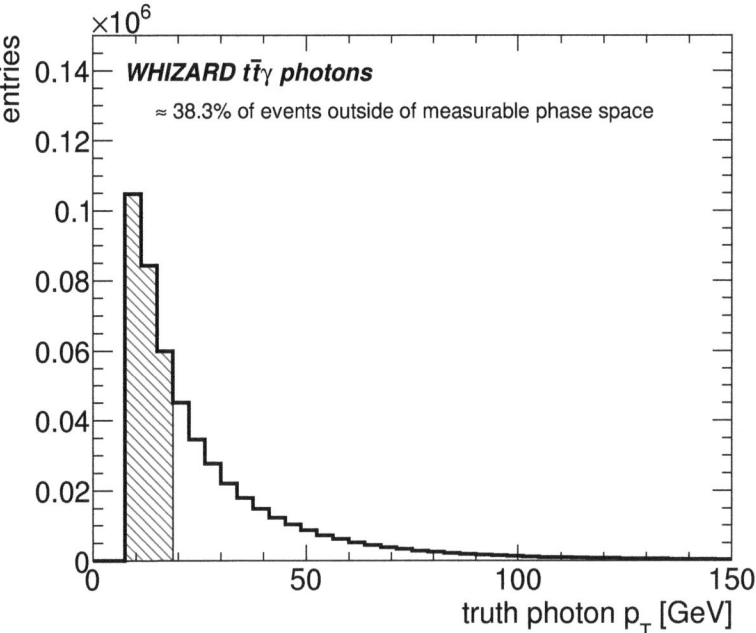

Figure 37: Distribution of the transverse momenta of truth photons generated by WHIZARD. The red shaded area below 15 GeV is excluded from the acceptance region, the measurement of the cross section is extrapolated to the full photon p_T spectrum.

Remark on the Calculation of the k-Factor Since the signal MC simulation contains a photon in the 7-particle final state whose ME comprises contributions from Feynman diagrams with $t\gamma$, $q\gamma$ and $\ell\gamma$ couplings, the calculated cross section strongly depends on the configuration of the phase space cuts used for the simulation. This is contrary to the standard $t\bar{t}$ production, where there are only two massive top quarks in the final state; hence there is no dependence of the total cross section on the considered phase space.

Since the magnitude of the k-factor for $t\bar{t}\gamma$ is severely affected by the choice of the considered phase space, a reliable comparison between the result of the cross section measurement and its theoretical prediction is only possible if the phase space cuts are carefully adjusted to the

detector acceptance (*fiducial cross section*). This topic is currently under investigation [123].
The current value of $k = 2.33 \pm 0.22$ assessed above is merely a rough estimation.

10.3 Background Samples

The background processes relevant for the $t\bar{t}\gamma$ cross section measurement are basically the same as for any other top quark pair analysis. The background contributions include W boson production with the additional emission of jets due to strong interaction (W+jets), Z+jets, single top production and the production of two EW bosons (di-boson processes WW, ZZ and WZ).
Multijet events solely produced via strong interaction cannot be modeled well and are therefore estimated from data (see Sec. 11.4.3).
For the W+jets background, two kinds of MC samples exist; once with and once without the explicit radiation of an additional photon. The W+jets sample without explicit photon production is used only for the comparison with data regarding the event preselection in Sec. 10.8. For the final cross section measurement, where the production of W+jets with an additional photon has to be considered, the W+jets+γ samples are used (see Sec. 11.4.2).

Top Quark Pair Production Top quark pair production without the explicit emission of a photon within the hard process was simulated with MC@NLO (Sec. 6.1) using the CTEQ6.6 PDF [124]. Parton showering, hadronization and the simulation of the underlying event (UE) was performed with HERWIG (Sec. 6.6.1) and JIMMY (Sec. 6.6.4). The description of photon radiation is improved by using PHOTOS (Sec. 6.6.3). The parameters of parton showering and the UE were set by the AUET1 tune [125].

The absolute cross section was determined to be $\sigma_{t\bar{t}} = 165^{+11}_{-16}\,\mathrm{pb}^{-1}$ by the HATHOR tool with an approximate NNLO calculation [126]. HERWIG radiates photons from the top quarks generated by MC@NLO as well as from all consecutive decay products due to Sudakov from factor approximation. Hence, there is an unwanted overlap between the 7-particle $t\bar{t}\gamma$ signal sample generated with WHIZARD and the MC@NLO $t\bar{t}$ sample. This overlap has been removed (see Sec. 10.4). In order to test various systematic uncertainties of the simulation of $t\bar{t}$ events, additional samples were created using the POWHEG and AcerMC MC generators:

POWHEG samples were produced with parton showering performed once with HERWIG and once with PYTHIA. The AcerMC $t\bar{t}$ samples were produced with different ISR and FSR settings for the parton showering with PYTHIA.

Single Top Quark Production Single top quark processes (see Sec. 2.3) were simulated with MC@NLO interfaced to HERWIG and JIMMY. Separate samples for s-channel, t-channel and associated production (Wt) were produced. For the Wt channel, there is an ambiguity with diagrams from $t\bar{t}$ production at NLO level. This overlap was removed by the *diagram-removal scheme* [127].

The cross sections were calculated at the approximate NNLO level and read $64.6^{+2.7}_{-2.0}\,\mathrm{pb}^{-1}$ for the t-channel, $4.6 \pm 0.2\,\mathrm{pb}^{-1}$ for s-channel and $15.7 \pm 1.1\,\mathrm{pb}^{-1}$ for the associated production mode [128–130].

$W \to \ell\nu$ with the Associated Production of Jets (W+Jets) Processes with the production of a single W boson and up to five partons in the final state were simulated with the ALPGEN (Sec. 6.3) generator using the CTEQ6L1 [131] PDFs. The simulated events were interfaced to HERWIG and JIMMY using the AUET1 tune for CTEQ6.1 [125].

Additional samples with a pair of heavy flavor quarks and up to three additional partons ($W + b\bar{b}$+jets, $W + c\bar{c}$+jets) were created in order to increase the relevant amount of events that will pass a requirement on b-tagged jets. The overlap of these additional samples with the standard W+jets samples is removed since they contain contributions from heavy quarks either [132].

$Z \to \ell^+\ell^-$ with the Associated Production of Jets (Z+Jets)
Processes with the production of a Z boson and up to five partons in the final state were simulated the same way as the W+jets samples using the ALPGEN generator interfaced to HERWIG and JIMMY with the AUET1 tune for CTEQ6.1 applied.
Also here, additional samples with explicit heavy quark content ($Z + b\bar{b}$+jets, $Z + c\bar{c}$+jets) were created and the overlap with the standard samples has been removed.

Di-Boson Production (WW, ZZ, WZ) Di-boson samples containing WW, WZ and ZZ events were generated with HERWIG. The bare cross sections calculated by HERWIG were corrected by k-factors obtained with the MCFM code [133].
The resulting cross sections are $1.48\,\text{pb}^{-1}$ for WW, $1.30\,\text{pb}^{-1}$ for WZ and $1.60\,\text{pb}^{-1}$ for ZZ production.

10.4 Signal Phase Space Overlap Removal

Since the simulated $t\bar{t}\gamma$ sample already includes photon radiation within the hard process, MC@NLO $t\bar{t}$ events with a photon radiated from the top quarks or their direct decay products simulated by HERWIG or PHOTOS can mimic $t\bar{t}\gamma$ signal events and hence constitute an unwanted overlap.

Events in the MC@NLO sample are checked if they contain a true photon on MC generator level that has been emitted by one of the top quarks or one of its direct decay products by investigating the process information provided by HERWIG. If this is the case, the invariant mass requirements of the WHIZARD sample are applied to that event. If the mass cuts are fulfilled and a true photon with $p_T^{\text{truth}} > 8\,\text{GeV}$ radiated from the HERWIG top quark decay exists, the event is removed from the event selection in order to avoid an overlap with the WHIZARD $t\bar{t}\gamma$ sample.

The remaining MC@NLO $t\bar{t}$ events can only contain true prompt photons that are outside the phase space defined by the invariant mass cuts and the minimum photon p_T according to the WHIZARD signal event generation (see Sec. 10.2). Hence such events have to be considered to be background.

First, the MC@NLO $t\bar{t}$ events before the overlap removal procedure are investigated if they survive the event selection due to a real physical $t\bar{t}\gamma$ event signature. This would happen if the detector acceptance and resolution covered a part of the $t\bar{t}\gamma$ phase space that is not included in the WHIZARD simulation.

The reconstruction of a true photon with $p_T^{\text{true}} < 8\,\text{GeV}$ as a selected photon with $p_T^{\text{reco}} \geq 15\,\text{GeV}$ is very unlikely.

The invariant mass cuts of 5 GeV between the truth photon and truth lepton/quarks could theoretically lead to angular distances between their associated reconstructed photons and jets/leptons large enough to enable both objects to be separated experimentally. Fig. 38 and Fig. 39 show the ΔR distributions between the truth photon radiated from top quarks or their direct decay products and the truth leptons/light quarks from the W boson decay (upper plots). The second plot in the figures depicts the ΔR distributions between the reconstructed photon and leptons/jets associated to their corresponding truth objects respectively.

The distributions are shown for different invariant mass cuts and indicate that lowering the requirement on the invariant masses would not increase the number of selected events and hence the choice of the cuts in the event selection is strict enough to avoid a leakage for $m_\text{inv} < 5\,\text{GeV}$ into the signal phase space defined for the WHIZARD $t\bar{t}\gamma$ sample.

Incongruities of Phase Space Coverage While WHIZARD calculates the full ME with outgoing quarks and the photon being generated simultaneously before the quarks are processed by HERWIG, photon radiation from quarks in the MC@NLO sample is performed later during the quark processing procedure. As consequence, photons radiated from quarks in MC@NLO cannot be consistently compared to photons simulated by the WHIZARD 7-particles final state.

In order to account for this inconsistency, four different scenarios in HERWIG have been compared: First, the quarks are considered as the direct decay products of the W boson (HERWIG status codes 123/124), then they undergo slight corrections due to the MC@NLO NLO subtraction method (HERWIG status codes 143/144). After fragmentation, which includes photon radiation, the quarks are assigned the "final state" status code 2 before they enter the hadronization process. Hence, the momenta of final state quarks with status code 2 are consistent with momentum conservation when considering the invariant mass between quark and photon.

Since also gluon radiation has already performed, which is not included in the WHIZARD sample, the quark momenta are recalculated by scanning through all parent particles of the quark with status code 2 and summing up the momenta of the child particles of these, excluding photons. By this way, the gluon momenta are recombined to the final quark momentum, hence the resulting momentum of this procedure can be considered to be closest to the WHIZARD simulation and has therefore been chosen to estimate the nominal expectation of the different $t\bar{t}$ background contributions.

10 SIGNAL AND BACKGROUND MODELING

Comparing all four possibilities to calculate the invariant masses between quarks and photons, no differences in the MC@NLO $t\bar{t}$ event yield could be observed after the signal overlap contributions had been removed.

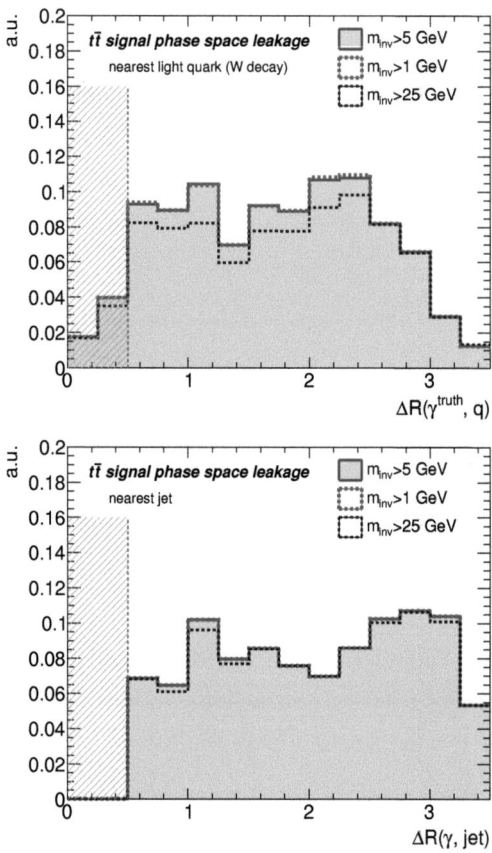

Figure 38: ΔR distributions between the truth photon radiated from the top quark and the nearest light quark originating from the W decay of the top quark (left plot) and the ΔR between the associated reconstructed photon and the jet nearest to the quark from the W decay, if existing (right plot); shown for the MC@NLO MC sample without any restrictions on the phase space. The cyan, solid histograms represent the nominal invariant mass cut of 5 GeV chosen for the signal event generation with the WHIZARD MC generator. The dashed red histograms show the expected ΔR distribution if the invariant mass cuts were lowered to 1 GeV, the green histograms demonstrate the effect of the choice of a large invariant mass cut of 25 GeV.

The comparison indicates that lowering the invariant mass cuts between photon and quarks (dashed red line) does introduce a slight additional phase space to the event selection which is however not present at reconstruction level anymore. The red dashed area depicts the distance in $\eta - \phi$ space between selected photons and jets excluded by the event selection (selection step 19, see Sec. 9.9).

10 SIGNAL AND BACKGROUND MODELING

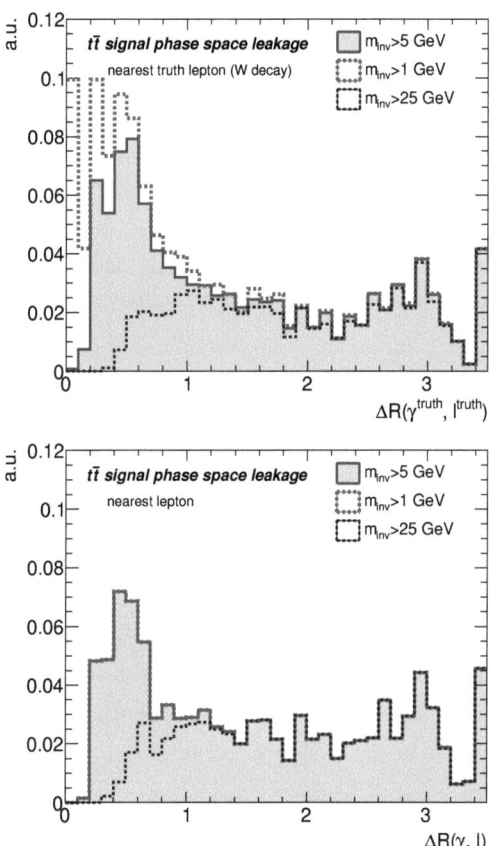

Figure 39: Compared to Fig. 38, a similar result can be observed for the ΔR distributions between photon and lepton. Here, the ΔR distributions between the truth photon and the truth lepton in the MC@NLO sample (left plot) and ΔR between the associated reconstructed photon and the selected lepton (right plot) are shown. While there seems to be a significant increase of events when lowering the invariant mass cuts on truth level, this behavior is not observed on reconstruction level anymore.

10.5 Pileup Reweighting

The MC samples used for analyses with data taken in 2011 were created in 2010, assuming a certain pileup configuration. At that time, the exact bunch spacing and luminosity conditions in the LHC could not be foreseen, hence the actual distribution of the average number of bunch crossings $\langle n_{\text{BX}} \rangle$ does not match the one applied on MC simulation [134].

Each event in the MC samples is therefore reweighted by the ratio w_{pileup} of relative frequency $\langle n_{\text{BX}}^{\text{Data}} \rangle$ w.r.t. the whole distribution $f(\langle n_{\text{BX}} \rangle)$ in data over the corresponding relative frequency $\langle n_{\text{BX}}^{\text{MC}} \rangle$ in MC simulation:

$$w_{\text{pileup}} = \frac{\int_{\langle n_{\text{BX,low edge}}^{\text{Data}} \rangle}^{\langle n_{\text{BX, up edge}}^{\text{Data}} \rangle} f_{\text{Data}}(\langle n_{\text{BX}} \rangle) \, d\langle n_{\text{BX}} \rangle}{\int f_{\text{Data}}(\langle n_{\text{BX}} \rangle) \, d\langle n_{\text{BX}} \rangle} \bigg/ \frac{\int_{\langle n_{\text{BX,low edge}}^{\text{MC}} \rangle}^{\langle n_{\text{BX, up edge}}^{\text{MC}} \rangle} f_{\text{MC}}(\langle n_{\text{BX}} \rangle) \, d\langle n_{\text{BX}} \rangle}{\int f_{\text{MC}}(\langle n_{\text{BX}} \rangle) \, d\langle n_{\text{BX}} \rangle} .$$
(136)

$\langle n_{\text{BX,low edge}} \rangle$ and $\langle n_{\text{BX,up edge}} \rangle$ are the lower and upper boundaries of the considered bin in the $\langle n_{\text{BX}} \rangle$ distributions and are given by the larger bin width of the MC distribution.

Fig. 40 shows a comparison between the pileup configuration in the MC samples and the actual $\langle n_{\text{BX}} \rangle$ distribution for data. Obviously, not all values of average bunch crossings simulated in the MC samples are actually present in data. Hence, some pileup weights are zero.

10.6 Event Weights

MC samples are generated with a fixed prediction concerning physics modeling on the one hand and a detector prescription on the other hand. As data analysis progresses, deviations between MC prediction and actual data will be observed. Based on the understanding of the sources of these deviations, detector and physics modeling will be adjusted afterwards in order to improve the matching between data and MC simulation.

10 SIGNAL AND BACKGROUND MODELING

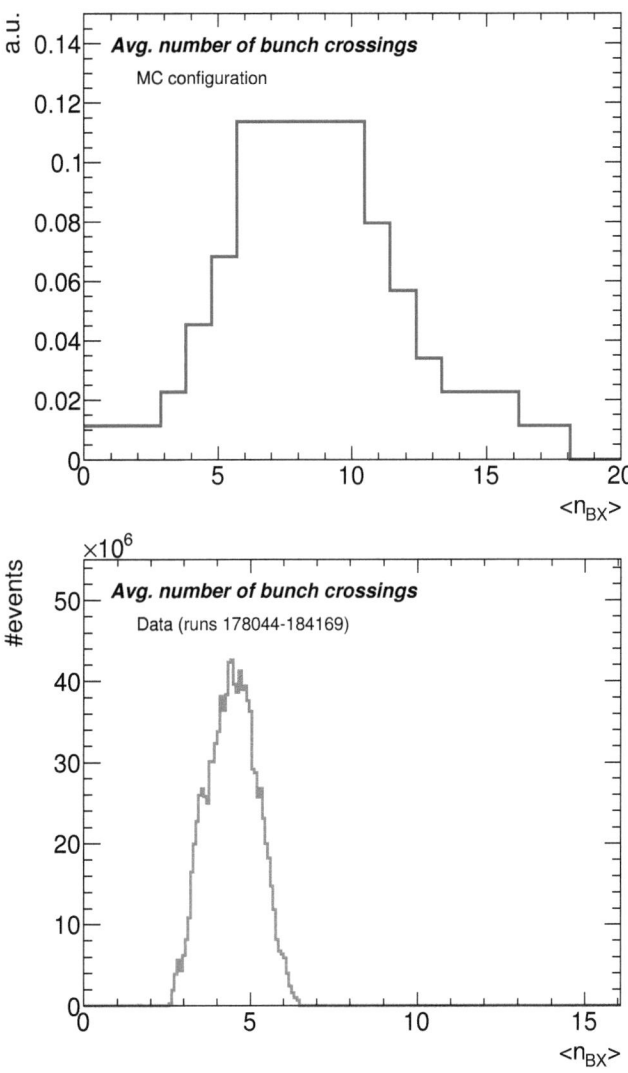

Figure 40: Distribution of average number of bunch crossings $\langle n_{\mathrm{BX}} \rangle$ for MC simulation (left plot) and data (right plot).

In order not to generate new MC samples regularly, which is very time-consuming, the existing MC samples are adapted to data expectation by introducing event weights which scale observables like kinematic variables or the expected event yield to the data expectation.

Basically, there are two kinds of event weights: Some are applied to each physical object in order to change their kinematic properties like momenta (*scale factors*), others change the expected yield of the absolute event normalization. All scale factors and weights are multiplied in order to obtain the final event weight.

On event level, two different weights exist: First, there is an MC weight which is generated by the chosen MC generator and is usually equal to one, except for MC@NLO, where $\approx 10\%$ of the events are applied a MC weight of -1. Secondly, the event is reweighted by the pile-up weight (Sec. 10.5).

10.6.1 Object Weights

There are various weights on object level (*scale factors* SF) that are multiplied on the event weight. Scale factors are defined as the ratio of an efficiency measured in data (ϵ_{Data}) over the same efficiency found in MC simulation (ϵ_{MC}):

$$\text{SF} = \frac{\epsilon_{\text{Data}}}{\epsilon_{\text{MC}}}. \tag{137}$$

Electron Scale Factors Electron trigger, reconstruction and identification efficiencies were measured from data using a tag and probe method for selected $Z \to ee$ events. The decay $J/\psi \to ee$ was investigated for electrons with a small transverse momentum [135]. Significant deviations of the electron ID efficiency had been observed in the low E_T region of $20\,\text{GeV} < E_T < 30\,\text{GeV}$. Hence, an additional E_T correction is applied on the electron ID SFs [136].

10 SIGNAL AND BACKGROUND MODELING

Muon Scale Factors Muon trigger, reconstruction and identification efficiencies had been measured using $Z \to \mu\mu$ decay [137].

b-Tagging Scale Factor b-tagged jets are weighted taking into account the b-tag efficiencies measured in data and found in MC simulations [90]. The efficiencies were determined for b-tagged, c-tagged and light quark jets for different values of jet $|\eta|$ and p_T. For each of these three jet types, scale factors had been evaluated by

$$\mathrm{SF}_{\mathrm{Flavor}} = \frac{\epsilon_{\mathrm{Flavor}}^{\mathrm{Data}}}{\epsilon_{\mathrm{Flavor}}^{\mathrm{MC}}}. \tag{138}$$

If the jet is tagged as a heavy quark flavor jet, it is applied a tag weight w_{jet} equal to the scale factor ($w_{\mathrm{jet}} = \mathrm{SF}_{\mathrm{Flavor}}$). Otherwise, the inefficiency is calculated by

$$w_{\mathrm{jet}} = \frac{1 - \epsilon_{\mathrm{Flavor}}^{\mathrm{Data}}}{1 - \epsilon_{\mathrm{Flavor}}^{\mathrm{MC}}}. \tag{139}$$

The product of jet weights $w = \prod w_{\mathrm{jet}}$ is multiplied on the event weight considering all selected jets.

10.6.2 Final Event Weight

Including the MC weight, the pileup weight and all object SFs, the overall event weight w_{event} finally reads

$$w_{\mathrm{event}} = w_{\mathrm{MC}} \times w_{\mathrm{pileup}} \times \prod_{\mathrm{sel.}\ e} (\mathrm{SF}_{\mathrm{trig.}}^{e} \cdot \mathrm{SF}_{\mathrm{reco.}}^{e} \cdot \mathrm{SF}_{\mathrm{ID}}^{e}) \times \prod_{\mathrm{sel.}\ \mu} (\mathrm{SF}_{\mathrm{trig.}}^{\mu} \cdot \mathrm{SF}_{\mathrm{reco.}}^{\mu} \cdot \mathrm{SF}_{\mathrm{ID}}^{\mu}) \times \prod_{\mathrm{sel.\ jets}} w_{\mathrm{jet}}. \tag{140}$$

The final event weight is applied to any event, regardless whether it has passed a certain step in the event selection or not. Other approaches consider the event weight only for selected events. A consistent, ATLAS-wide approach had been under discussion when this thesis was created.

10.7 MC Uncertainties

The uncertainties of MC predictions are estimated regarding $t\bar{t}$ MC simulation, W+jets, Z+jets and QCD (multijet) contributions.

Considering $t\bar{t}$ simulation, the effects of the usage of different showering algorithms (HERWIG vs. PYTHIA), and two different MC generators at the NLO level (MC@NLO vs. POWHEG) have been studied. On LO level, various MC samples generated with AcerMC (Sec. 6.4) have been provided by the ATLAS top quark physics group where different settings for the inclusion of ISR and FSR have been applied. The overall uncertainty for the $t\bar{t}$ and the $t\bar{t}\gamma$ samples yields $^{+11.7}_{-9.1}$ %. A breakdown of the several uncertainties that sum up to that number is presented in Tab. 18, a detailed description of the evaluation of the $t\bar{t}$ uncertainties is given in Sec. 12.3.7.

Tab. 9 provides a complete list of expected event yields for the various $t\bar{t}$ MC samples as reference. Note that the numbers given there are scaled with the event weight established in Sec. 10.6.2 whereas the uncertainty is evaluated based on the bare MC weights in order to disentangle uncertainties of the event weights from the calculation of the MC uncertainties.

For the W+jets MC samples, the uncertainty of the Berends-Giele scaling [138, 139] is considered as well as the uncertainty on the additional production of W+jets with explicit heavy quark flavor decays $(pp \rightarrow W + b\bar{b} + Np\,(p = 0\ldots 3)$, $pp \rightarrow W + c\bar{c} + Np\,(p = 0\ldots 3)$, $pp \rightarrow W + c(\bar{c}) + Np\,(p = 0\ldots 4))$. The Berends-Giele scaling assumes an uncertainty of 24 % for each additional simulated jet (parton). Thus, events from $W + 1p$ samples are applied an uncertainty of 24 %, those from $W + 4p$ samples an uncertainty of 48 %. In this scenario, $W + b\bar{b}(c\bar{c}) + 0p$ samples have an uncertainty of $\sigma_W = \sqrt{2 \cdot 24\,\%}$.

The uncertainty due to Berends-Giele scaling has been applied to Z+jets samples either, treating additional, explicit heavy quark flavors the same way as for the W+jets samples.

The contributions of the heavy quark flavor samples are multiplied with an additional scale factor of 1.63 ± 0.76 for $W+b\bar{b}(c\bar{c})$ and 1.11 ± 0.35 for $W+c(\bar{c})$ respectively, taking their related uncertainties into account. This additional scaling is not applied to Z+jets samples. The background from multijet events has been estimated using a data-driven technique (see Sec. 11.4.3) since QCD processes cannot be modeled precisely in MC simulations. An uncertainty of 50 % is applied to this estimation before the b-tag selection and a 100 % uncertainty if ≥ 1 b-tagged jets are required by the event selection, respectively.

10.8 Event Yields

The MC samples have been produced for a certain integrated luminosity L_{MC}. In order to obtain a MC event yield for a given integrated luminosity L of the considered data, the event yields have to be scaled by $s_L = \frac{L}{L_{\text{MC}}}$. Given $N_{\text{all}} = L_{\text{MC}} \cdot \sigma \cdot k$, the scaling yields the normalized expected number of events N_{exp} for the given luminosity $L = 1.04 \, \text{fb}^{-1}$ via

$$N_{\text{exp}} = s_L \cdot \tilde{N} = \frac{\tilde{N}}{\tilde{N}_{\text{all}}} \cdot L \cdot \sigma \cdot k \,, \tag{141}$$

where $\tilde{N} = \sum_{i=1}^{N} w_{\text{event},i}$ is the scaled number of MC events that pass the event selection, σ is the predicted cross section of the simulated process and k its k-factor. $\tilde{N}_{\text{all}} = \sum_{i=1}^{N_{\text{all}}} w_{\text{event},i}$ is the overall weighted number of events contained in the considered MC sample.

The event preselection is applied to data and to all MC simulations. Fig. 41 to 44 show the comparison between data and simulation before the requirement of a b-tagged jet (after event selection step 14), Fig. 45 to Fig. 48 illustrate the corresponding distributions after the requirement of a b-tagged jet (after event selection step 15) (see Sec. 9.9).

Fig. 41 and Fig. 45 depict the transverse momentum of all selected jets, Fig. 42 and Fig. 46 the p_T of the selected lepton ($p_T(e)$ for the

e+jets channel and $p_T(\mu)$ for the μ+jets channel respectively). The missing transverse energy $\slashed{E}_T$ is shown in 43 and 47, the transverse W mass $M_T(W)$ in 44 and 48. The results are presented for the e+jets channel and the μ+jets channel.

The corresponding breakdown of the event yields for the signal and background processes compared with the corresponding event yields of data are presented in Tab. 8. A detailed overview of the composition of the Z+jets, W+jets, di-boson and single top quark contributions is presented in App. A.

Comparing Fig. 41 to Fig. 44 with Fig. 45 to Fig. 48, the large contribution of uncertainties due to the Berends-Giele scaling on the W+jets normalization can be observed. The relative amount of events including a top quark is significantly increased by a b-jet requirement. Generally, the data yields and the predictions from MC simulation are in good agreement, indicating that all relevant physical processes have been taken into account.

10.8.1 Full Event Selection and Signal Efficiency

Applying the full event selection including the photon requirement yields 52 events in the e+jets channel and 70 events in the μ+jets channel respectively. The $t\bar{t}\gamma$ signal expectation is $20.1 \pm 0.3 \pm 3.8$ ($26.5 \pm 0.3 \pm 5.0$) in the e+jets (μ+jets) channel for the quoted cross section of $\sigma_{t\bar{t}\gamma}^{MC} = 840\,\text{fb}$ and the estimation of the k-factor of $k_{t\bar{t}\gamma} = 2.33 \pm 0.22$ (see Sec. 10.2).

The event selection efficiency for $t\bar{t}\gamma$ signal events is $[0.992 \pm 0.014(\text{stat.}) \pm 0.187(\text{syst.})]$ % in the e+jets channel and $[1.31 \pm 0.02(\text{stat.}) \pm 0.2(\text{syst.})]$ % in the μ+jets channel respectively. The quoted systematic uncertainty contains the $t\bar{t}$ MC modeling uncertainties (see Sec. 10.7) as well as the uncertainty due to the estimation of the k-factor.

10 SIGNAL AND BACKGROUND MODELING

	before b-tagging (pre-tag)		after b-tagging	
	e+jets	μ+jets	e+jets	μ+jets
$t\bar{t}\gamma$	98 ± 9	131 ± 12	86 ± 8	115 ± 11
$t\bar{t}$	4700 ± 600	6800 ± 700	4100 ± 560	6000 ± 600
W+jets	5900 ± 1000	10800 ± 1700	850 ± 150	1600 ± 300
Z+jets	650 ± 110	870 ± 130	95 ± 15	120 ± 15
Di-boson	78 ± 3	123 ± 5	13 ± 1	21 ± 1
Single top	300 ± 15	410 ± 20	240 ± 10	320 ± 15
Multijet	1000 ± 500	1460 ± 730	300 ± 300	500 ± 500
Sum	12650 ± 1250	20600 ± 2000	5700 ± 600	8700 ± 800
Data	11900 ± 100	19050 ± 140	5750 ± 80	8900 ± 90

Table 8: Comparison of event yields after the event preselection between data and the several signal and background contributions. The numbers are given before and after the b-tag requirement. The quoted uncertainties are described in Sec. 10.7, the uncertainties of data are the statistical ones. The sum of the predictions and the data yield are in good agreement.

	before b-tagging (pre-tag)		after b-tagging	
	e+jets	μ+jets	e+jets	μ+jets
MC@NLO	4656 ± 5	6788 ± 6	4104 ± 5	5996 ± 6
POWHEG (HERWIG)	5019 ± 12	7222 ± 15	4410 ± 10	6350 ± 14
POWHEG (Pythia)	4845 ± 12	7000 ± 15	4290 ± 10	6210 ± 14
AcerMC (nominal)	5170 ± 20	7290 ± 30	4590 ± 20	6500 ± 25
AcerMC (ISR up)	5490 ± 20	7720 ± 30	4860 ± 20	6840 ± 25
AcerMC (ISR down)	4640 ± 20	6700 ± 25	4140 ± 20	5970 ± 25
AcerMC (FSR up)	4990 ± 20	7010 ± 25	4380 ± 20	6180 ± 25
AcerMC (FSR down)	5350 ± 20	7630 ± 30	4780 ± 20	6800 ± 25
AcerMC (ISR+FSR up)	5180 ± 20	7450 ± 30	4540 ± 20	6550 ± 25
AcerMC (ISR+FSR down)	4880 ± 20	6980 ± 30	4360 ± 20	6250 ± 25

Table 9: Comparison of event yields of the various $t\bar{t}$ MC samples after the event preselection. The numbers are given before and after the b-tag requirement. Here, only the statistical uncertainties are quoted.

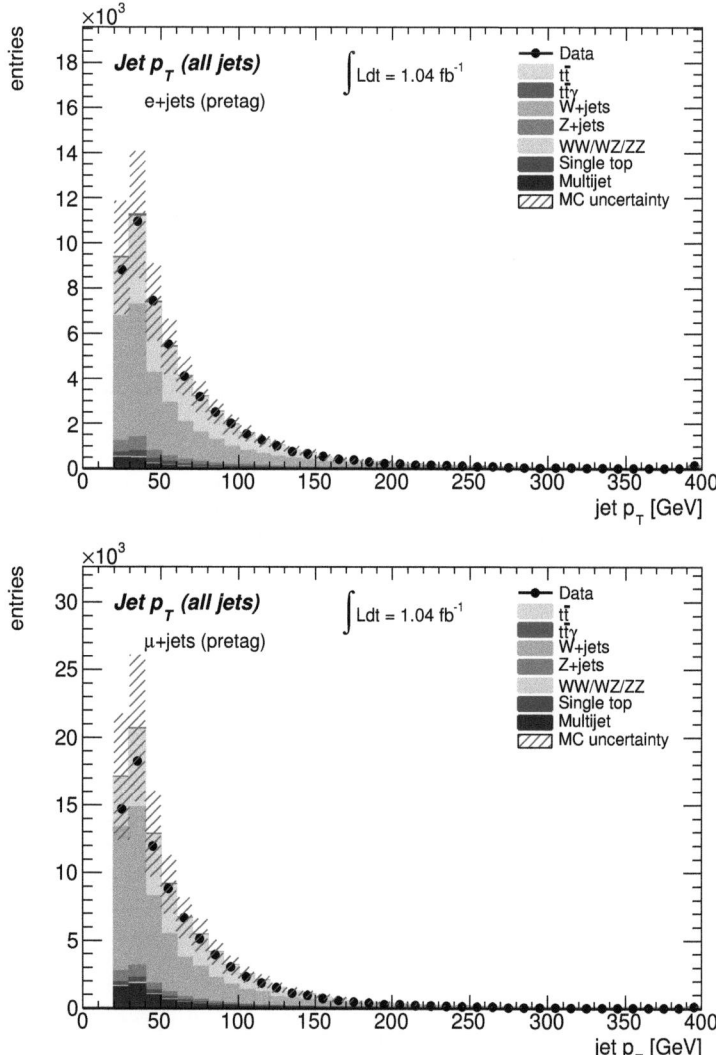

Figure 41: Jet p_T comparison between data and MC simulation for the event preselection before the b-tag requirement in the e+jets channel (upper plot) and the μ+jets channel (lower plot).

10 SIGNAL AND BACKGROUND MODELING 163

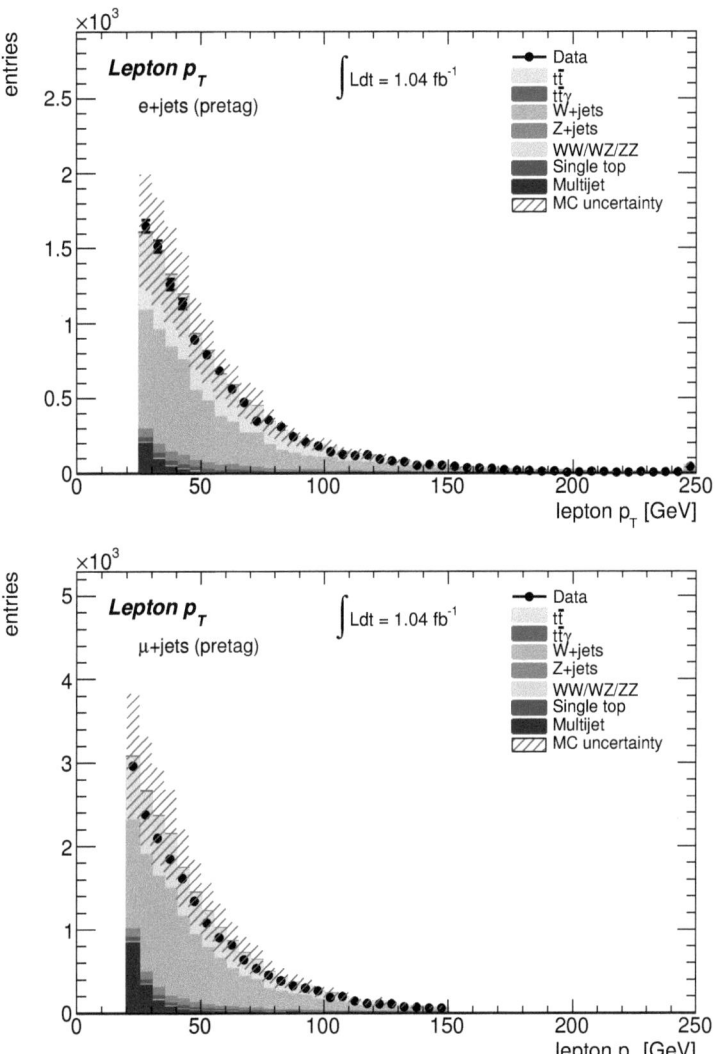

Figure 42: Lepton p_T comparison between data and MC simulation for the event preselection before the b-tag requirement in the e+jets channel (upper plot) and the μ+jets channel (lower plot).

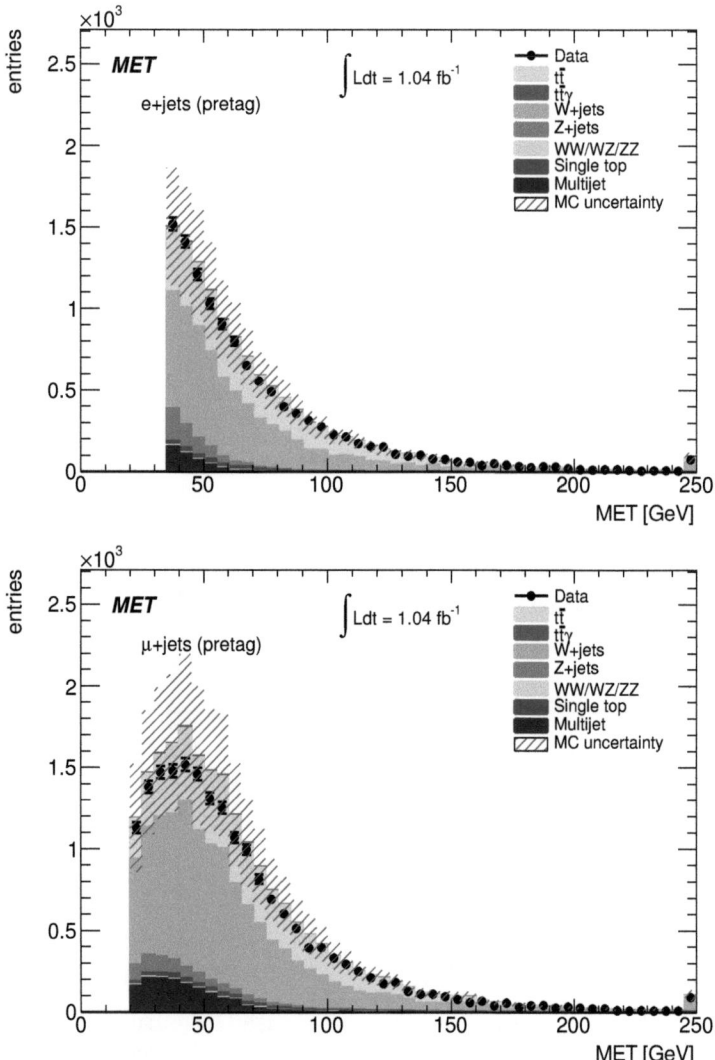

Figure 43: MET comparison between data and MC simulation for the event preselection before the b-tag requirement in the e+jets channel (upper plot) and the μ+jets channel (lower plot).

10 SIGNAL AND BACKGROUND MODELING

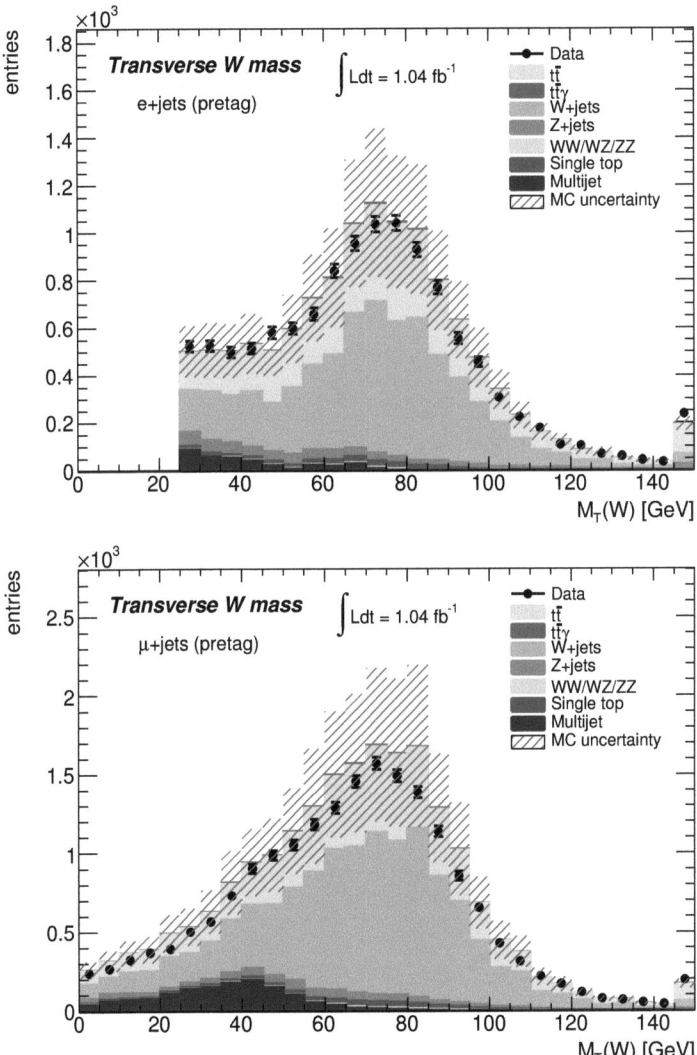

Figure 44: Transverse W mass comparison between data and MC simulation for the event preselection before the b-tag requirement in the e+jets channel (upper plot) and the μ+jets channel (lower plot).

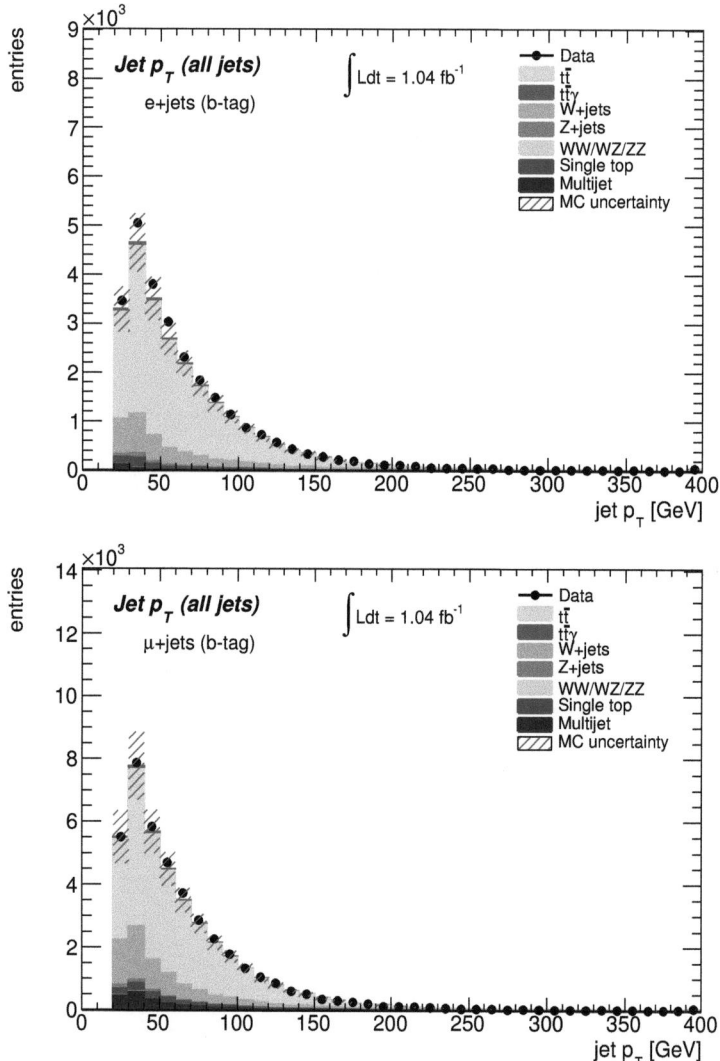

Figure 45: Jet p_T comparison between data and MC simulation for the event selection after the b-tag requirement in the e+jets channel (upper plot) and the μ+jets channel (lower plot).

10 SIGNAL AND BACKGROUND MODELING

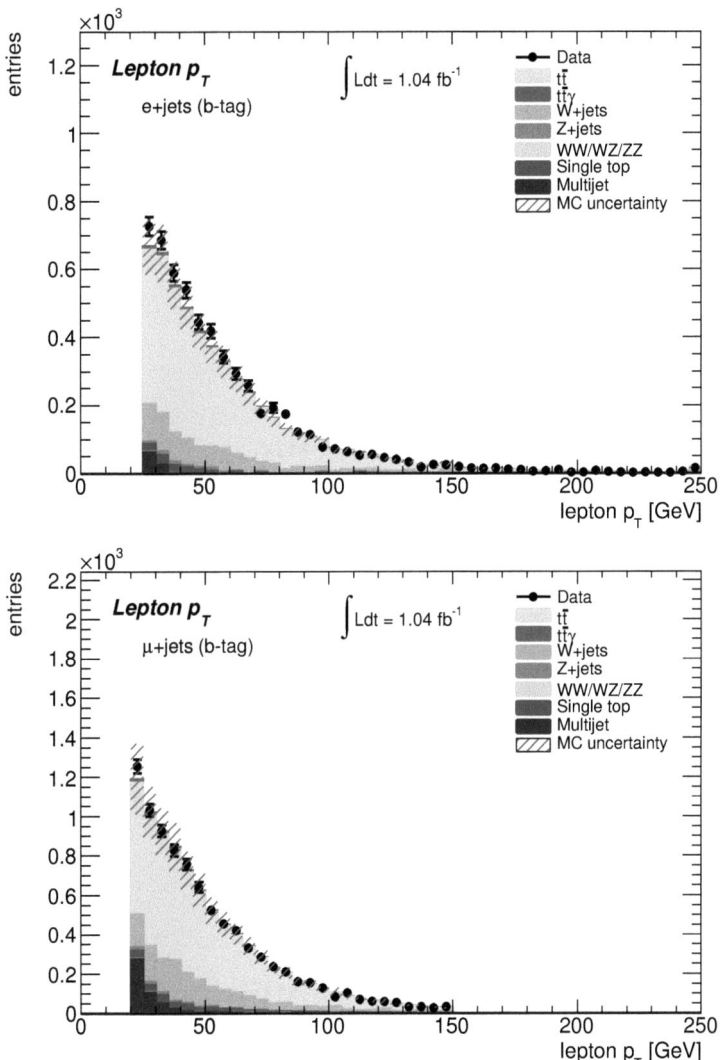

Figure 46: Lepton p_T comparison between data and MC simulation for the event selection after the b-tag requirement in the e+jets channel (upper plot) and the μ+jets channel (lower plot).

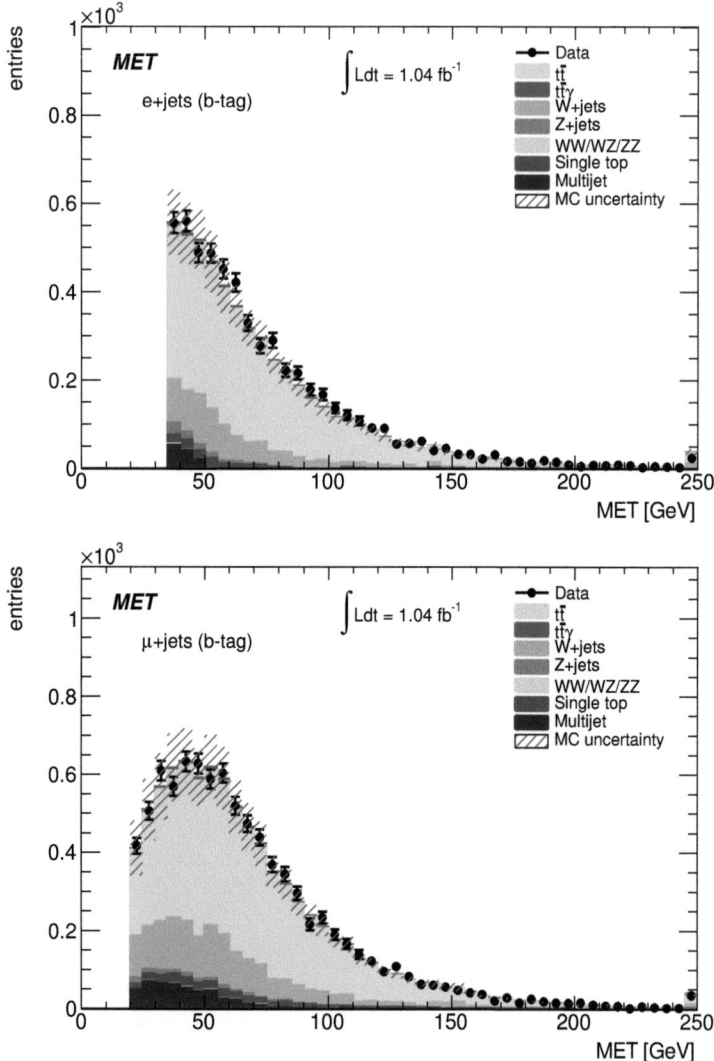

Figure 47: MET comparison between data and MC simulation for the event selection after the b-tag requirement in the e+jets channel (upper plot) and the μ+jets channel (lower plot).

10 SIGNAL AND BACKGROUND MODELING

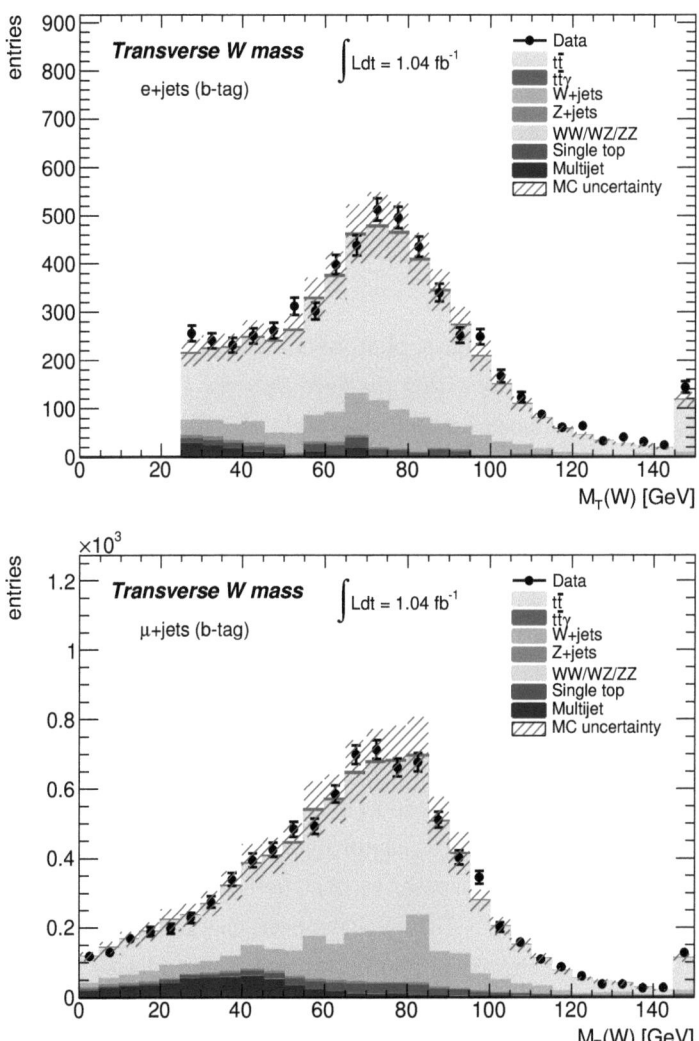

Figure 48: Transverse W mass comparison between data and MC simulation for the event selection after the b-tag requirement in the e+jets channel (upper plot) and the μ+jets channel (lower plot).

11 Cross Section Measurement

The central challenge of the measurement of the $t\bar{t}\gamma$ cross section is the distinction between photons from the signal process and all other photons. Basically, photons radiated from any physical process can have two different origins:

Prompt Photons Prompt photons are photons that are radiated before jet fragmentation within the hard process. Physically, this kind of photons is produced on the time scale of hadronization, i.e. not later than $\approx 10^{-24}$ s after the primary interaction. Prompt photons incorporate the photons produced in the $t\bar{t}\gamma$ signal process.

Photons from hadronic Decays ("Hadron Fakes") Light neutral mesons like π^0, η or ρ produced during hadronization can decay into two or three photons (*Dalitz decays*). Besides, also baryons and other (charged and excited) mesons can decay into photons.

Fig. 49 shows the composition of sources of such hadron fakes, extracted from HERWIG decay chain information in the MC@NLO $t\bar{t}$ MC sample. For this purpose, reconstructed photons with a transverse momentum of $p_T > 15$ GeV have been checked if they match to a truth photon. If this truth photon has not been radiated within the hard process, its origin has been determined from the HERWIG decay chain information. Dalitz decays from the process $\pi^0 \to \gamma\gamma$ make up the dominant fraction of hadron fakes.

Radiation from leptons occurs if the leptons are produced by the decay of a meson or a baryon if their energy is above the threshold where photon radiation is considered. A further branching of $\ell \to \ell\gamma$ is then taken into account.

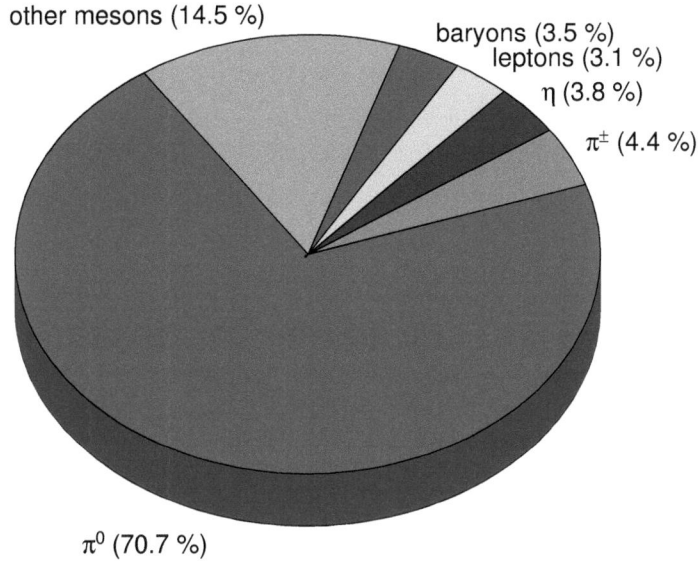

Figure 49: Sources of hadron fakes, as modeled by the HERWIG hadronization and decay algorithms.

11.1 Identification of Hadron Fakes

Hadronization is modeled in PS programs only approximately and is based on empiric expectations how hadrons are produced during jet fragmentation and how they decay. Furthermore, the shower shapes in the EMC cannot be simulated at a high precision due to the imperfect knowledge of the behavior and distribution of the material in the detector. Hence, the isolation shape of hadron fakes cannot be estimated from MC simulation but has to be extracted from data.

In order to distinguish between prompt photons and hadron fakes reliably, the tight photon definition (see Sec. 7.4.2) is modified by reverting the requirements on the shower shape variables sensitive to decays of neutral mesons, such as π^0, ρ or η, into a pair of close-by photons. The discriminating variables are (see Tab. 5) ΔE, F_{side}, $w_{s,3}$ and E_{ratio} which discriminate real photons from hadron decays in the fine segmented strip layer of the EMC. The photon definition is reverted by a logical inversion, i. e. at least one of these four requirements must not be fulfilled.

All other properties of the nominal photon definition, like the minimum transverse momentum and the restriction on the pseudo-rapidity, are kept.

11.2 Choice of the Isolation Discriminator

The main difference between prompt photons and hadron fakes is their difference in isolation: While prompt photons are usually well isolated, hadron fakes are produced inside a jet and are therefore surrounded by many tracks from charged mesons, baryons and leptons; or, in the case of calorimeter isolation, surrounded by additional cluster cells. Hence, hadron fakes are usually poorly isolated.

The effect of the choice of different photon isolation variables has been studied by comparing the selection efficiency of signal photons from the $t\bar{t}\gamma$ simulation with the simultaneous rejection of background photons obtained from a di-jet MC sample (JF17, see Tab. 32) by iteratively fixing a minimum value of isolation. Events having at least one good photon with $p_T \geq 15\,\text{GeV}$ and at least one primary vertex with a minimum of five tracks associated to it are selected. Events that contain a photon reconstructed within problematic detector regions due to the LAr hardware failure are rejected.

The efficiency of the photon selection is obtained by dividing the integral of the isolation distribution up to the chosen isolation cut $p_{T,\text{iso}}^{\text{cut}}$ over the whole integral:

$$\epsilon = \frac{\int_{\text{min}}^{p_{T,\text{iso}}^{\text{cut}}} f(p_{T,\text{iso}}) dp_{T,\text{iso}}}{\int_{\text{min}}^{\text{max}} f(p_{T,\text{iso}}) dp_{T,\text{iso}}}.$$

Fig. 50 compares the isolation shapes for one calorimeter isolation (half cone size $R = 0.2$) and one track isolation (half cone size $R = 0.2$). The isolation variables of prompt photons tend to have smaller values than those of background photons and the distributions are generally more narrow.

The behavior of photon selection efficiency vs. background photon rejection, which is defined as $1 - \epsilon_{\text{bkgd}}$, is studied for two calorimeter isolations with a half cone size of $R = 0.2$ and $R = 0.3$ and for two track isolations with the same half cone sizes. Fig. 51 indicates that the choice of $p_T^{\text{iso,cone20}}$ has the largest discriminating power since it provides a higher background rejection for a given signal efficiency.

11.3 Modeling of Signal and Background Contributions

In order to exploit the difference between the shapes of the isolation of prompt photons and photons from hadronic decays, the measurement of the inclusive $t\bar{t}\gamma$ cross section is performed by a template fit method using the $p_{T,\text{iso}}^{\text{cone20}}$ track isolation.

Photons remaining after the full $t\bar{t}\gamma$ event selection contain both prompt photons and hadron fakes. The contribution from hadron fakes is fully covered by fitting a hadron fake template as described in Sec. 11.3.3. The remaining prompt photons can be split into an amount of signal and contributions from various background sources. The yield of background contributions is estimated partially from data and from MC simulation. For any background estimation, only the contribution of prompt photons is considered since its hadron fake contribution is already included in the overall hadron fake estimation.

While the estimator for the signal contribution and the contribution of hadron fakes is treated as free, uniform parameter in the template fit, the contributions of the prompt photon backgrounds are assigned a

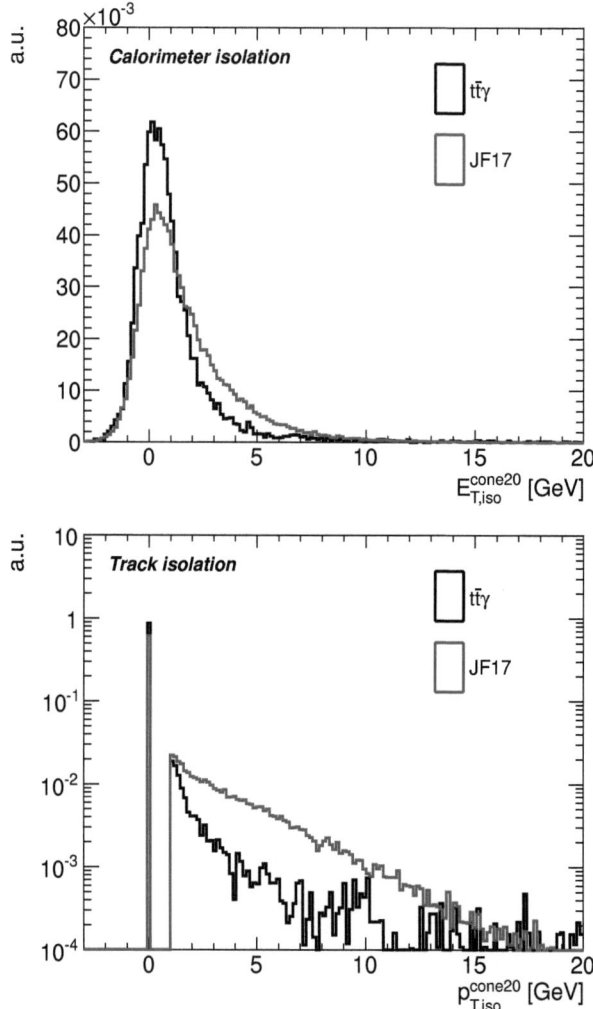

Figure 50: Comparison of isolation shapes between the $t\bar{t}\gamma$ signal MC sample and the JF17 di-jet background sample, shown for two different isolation variables; once with calorimeter isolation for a $R = 0.2$ half cone ($E_{\text{T,iso}}^{\text{cone20}}$, upper plot) and once for the track isolation with a half cone size of $R = 0.2$ ($p_{\text{T,iso}}^{\text{cone20}}$, lower plot). Note that the track isolation is calculated only for tracks with $p_\text{T} > 1$ GeV. If there is no track above this threshold, the track isolation is set to zero.

11 CROSS SECTION MEASUREMENT

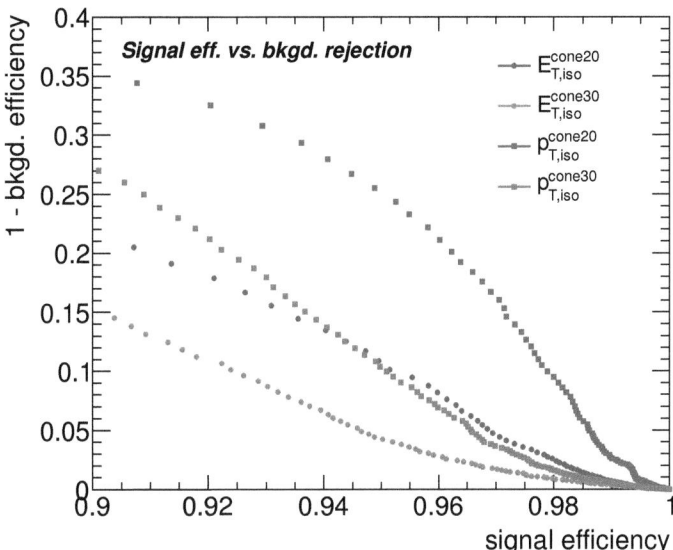

Figure 51: Background photon rejection vs. selection efficiency for two calorimeter isolations with cone size of $R = 0.2$ and $R = 0.3$ and for two track isolations with the same half cone sizes. The track isolations provide a higher background rejection for a given signal efficiency. The error bars have been omitted for a better legibility, all statistical uncertainties are at the order of $\approx 1\,\%$.

prior probability assuming a Gaussian distribution to take into account the statistical uncertainties of the corresponding background yield.

11.3.1 Template Fit

Template fits represent the estimation of a parameter by maximizing a binned log-likelihood for one or more models provided as histograms. For a given number N_{bins} bins of a template and a number $n_{\text{bkgd.}}$ of background sources, the likelihood L is given by

$$L = \prod_{i=1}^{N_{\text{bins}}} P(N_i | \lambda_i) \prod_{j=1}^{N_{\text{bkgd.}}} P(b_j) \cdot P(s)$$

where $P(N_i|\lambda_i)$ is the probability to find N_i entries in the i-th bin for a given estimation λ_i of entries; $P(b_j)$ and $P(s)$ are the prior probabilities for background contribution b_j and for signal respectively. Each bin content λ_i is the sum of the signal and overall background estimation for bin i:

$$\lambda_i = s_i + \sum_{j=1}^{N_{\text{bkgd.}}} b_{i,j}\,.$$

Although the background contributions are estimated in the e+jets and the μ+jets channel separately, the signal expectation s can be combined in one template fit:

$$P(N_i|\lambda_i) \to P(N_i^{e+\text{jets}}|\lambda_i^{e+\text{jets}}) \cdot P(N_i^{\mu+\text{jets}}|\lambda_i^{\mu+\text{jets}})\,,$$
$$P(b_j) \to P(b_j^{e+\text{jets}}) \cdot P(b_j^{\mu+\text{jets}})\,,$$
$$\lambda_i \to \epsilon_i \cdot s + \sum_{j=1}^{N_{\text{bkgd..}}} b_{i,j}\,,$$

where ϵ_i is the signal efficiency in bin i which is given by the $t\bar{t}\gamma$ event selection efficiency.

The template fit is performed with the Bayesian Analysis Toolkit (BAT package) [140]. A complete list of prior types applied on signal and the several background contributions is presented in Tab. 17.

11.3.2 Prompt Photon Template

The template for prompt photons is created from data considering $Z \to ee$ decays. This is possible since electrons and photons are reconstructed with the same algorithm and prompt electrons and prompt photons are believed not to differ very much in their isolation distributions[18].

---
[18] At the time the analysis in this thesis was performed, no dedicated photon sample extracted from data was available.

11 CROSS SECTION MEASUREMENT

A data sample enriched in $Z \to ee$ decays is obtained by requiring two tight electrons, one with a minimum transverse momentum of $p_T > 25\,\text{GeV}$ and the second one with a transverse momentum of $p_T > 15\,\text{GeV}$. The electron trigger has had to be fired.
The electron with the higher transverse momentum is required to match the corresponding trigger object; the isolation for the template is solely extracted from the electron with the smaller transverse momentum in order to avoid any trigger bias.
The invariant mass of the two selected electrons is required to be in the range between $66\,\text{GeV} \leq M_{\text{inv}} \leq 106\,\text{GeV}$ (mass window of the Z boson) in order to reduce multijet background. Fig. 52 shows the comparison of the invariant two-electron mass between data and $Z \to ee$ MC simulation, indicating that there is a slight increase of multijet background for smaller invariant mass values.
Since top quark events exhibit a larger number of hadronic jets compared to $Z \to \ell^+\ell^-$ production, the isolation of electrons of the latter process is generally slightly better than that for prompt photons originating from top quark processes. In order to account for the different event topologies, a small correction Δs_{MC}, derived from MC simulations, is applied to the isolation distribution of data bin per bin:

$$s_\gamma^{\text{Data}} = s_e^{\text{Data}} + \Delta s^{\text{MC}} \quad \text{with} \quad \Delta s^{\text{MC}} = s_\gamma^{\text{MC}} - s_e^{\text{MC}}. \tag{142}$$

Δs^{MC} is calculated as the difference of corresponding bin contents of normalized isolation distributions: s_γ^{MC} is taken from selected photons ($p_T \geq 15\,\text{GeV}$) of the $t\bar{t}\gamma$ sample, s_e^{MC} from the $Z \to ee$ MC samples (see Tab. 38).
Fig. 53 shows the comparison between the electron isolation obtained from $Z \to ee$ MC simulation and photon isolation taken from the $t\bar{t}\gamma$ MC sample.

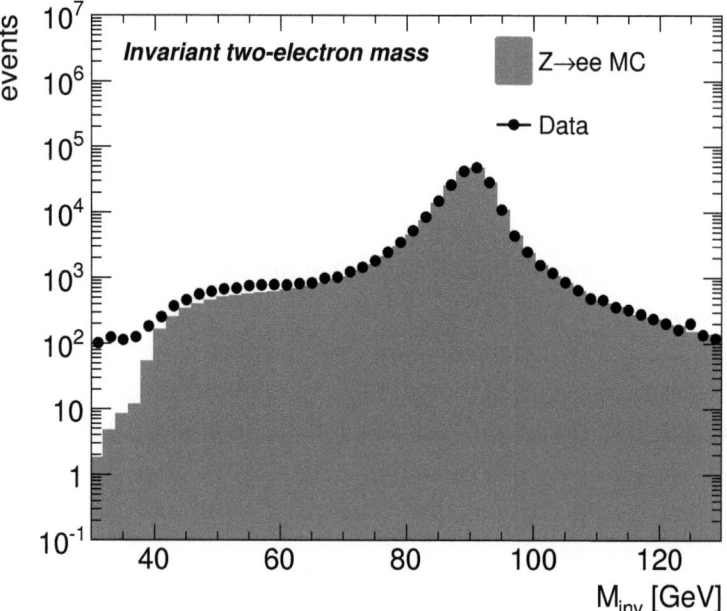

Figure 52: Comparison of the invariant the two-electron mass between data and $Z \to ee$ MC simulation. For small invariant masses, an increasing, slight discrepancy between data and MC simulation due to multijet background processes becomes visible.

11 CROSS SECTION MEASUREMENT

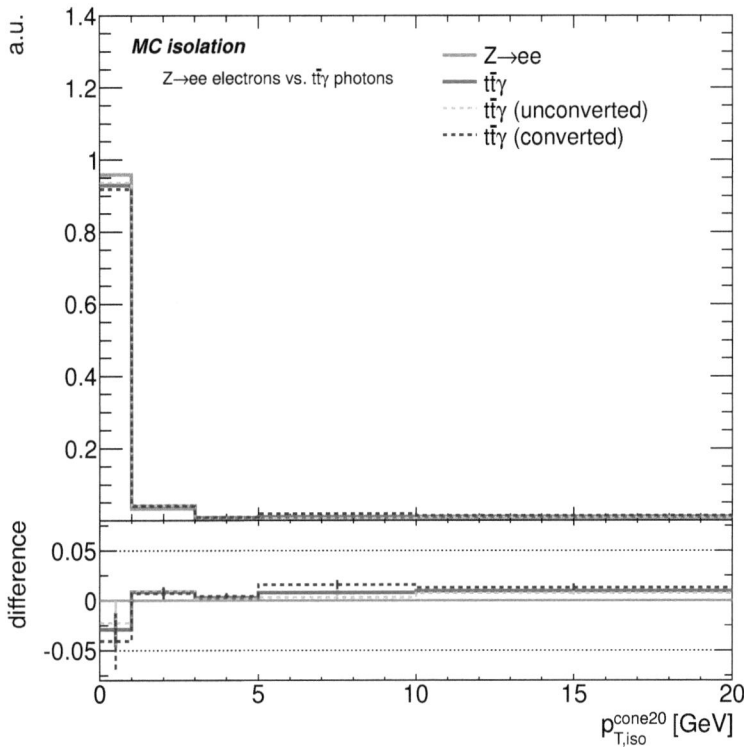

Figure 53: Comparison between the electron isolation obtained from $Z \to ee$ MC simulation and photon isolation taken from the $t\bar{t}\gamma$ MC sample. Δs^{MC} in Eq. (142) is calculated as the difference of corresponding bin contents between $t\bar{t}\gamma$ photons (solid red histogram) and $Z \to ee$ electrons (solid green histogram). The dashed histograms show the isolation of $t\bar{t}\gamma$ photons for converted and unconverted photons separately, indicating that there is no significant dependence of Δs^{MC} on the type of photons.

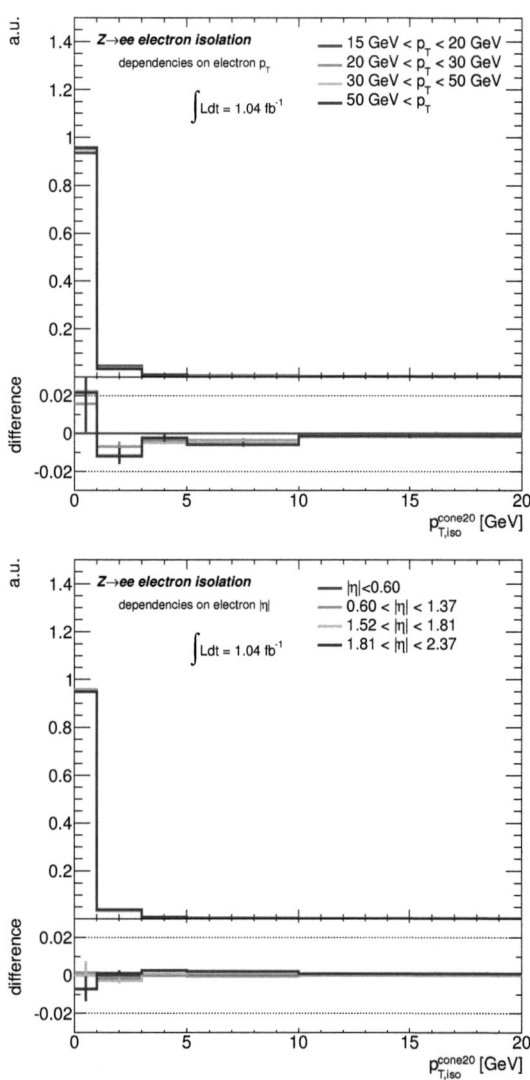

Figure 54: Comparison between the shapes of electron track isolation of $Z \to ee$ data events for different bins in p_T (upper plot) and $|\eta|$ (lower plot). The plots indicate that there is no dependency on any of these kinematic quantities within statistical uncertainties.

The isolation distributions are additionally created for four bins in p_T and four bins in $|\eta|$ in order to investigate possible dependencies of the isolation shapes on the electron (photon) kinematics and detector geometry. The distributions are divided into five bins in $p_\mathrm{T,iso}^\mathrm{cone20}$: $[0, 1)\,\mathrm{GeV}$, $[1, 3)\,\mathrm{GeV}$, $[3, 5)\,\mathrm{GeV}$, $[5, 10)\,\mathrm{GeV}$ and $[10, \infty)\,\mathrm{GeV}$

Fig. 54 shows the electron isolation as a function of p_T (left plot) and $|\eta|$ (right plot). The histograms indicate that there is no dependence of the isolation on kinematic quantities within statistical uncertainties (indicated by the pull plot in the lower panes), hence the overall prompt photon isolation distribution can be used as template. Values of $p_\mathrm{T,iso}^\mathrm{cone20} > 20.0\,\mathrm{GeV}$ are included as overflows in the last bin.

11.3.3 Hadron Fake Template

The hadron fake template is created from data using the jet/$\not{E}_\mathrm{T}$ data stream (see Sec. 8.1). Events exhibiting at least one primary vertex with a minimum of five associated tracks and at least one photon tagged as a hadron fake (see Sec. 11.1) with a transverse momentum of $p_\mathrm{T} \geq 15\,\mathrm{GeV}$ are selected. Triggers are not required.

Hadron fake templates are produced in four bins of transverse momentum and four bins of pseudo-rapidity $|\eta|$ in order to investigate dependencies of the track isolation of the hadron fakes on kinematic quantities. Fig. 55 indicates that the isolation spectrum of hadron fakes is depending significantly on the p_T of the photon. While for $|\eta| < 1.81$ the isolation shapes are not strongly depending on pseudo-rapidity, a large deviation can be observed for $|\eta| > 1.81$. Since hadron fakes are produced inside jets, they will be radiated more collinear with increasing momentum, thus being surrounded by an increasing number of tracks from the hadronic components of the jet. Similarly, hadron fakes with a large $|\eta|$ are more likely caved inside the jet due to a higher Lorentz boost.

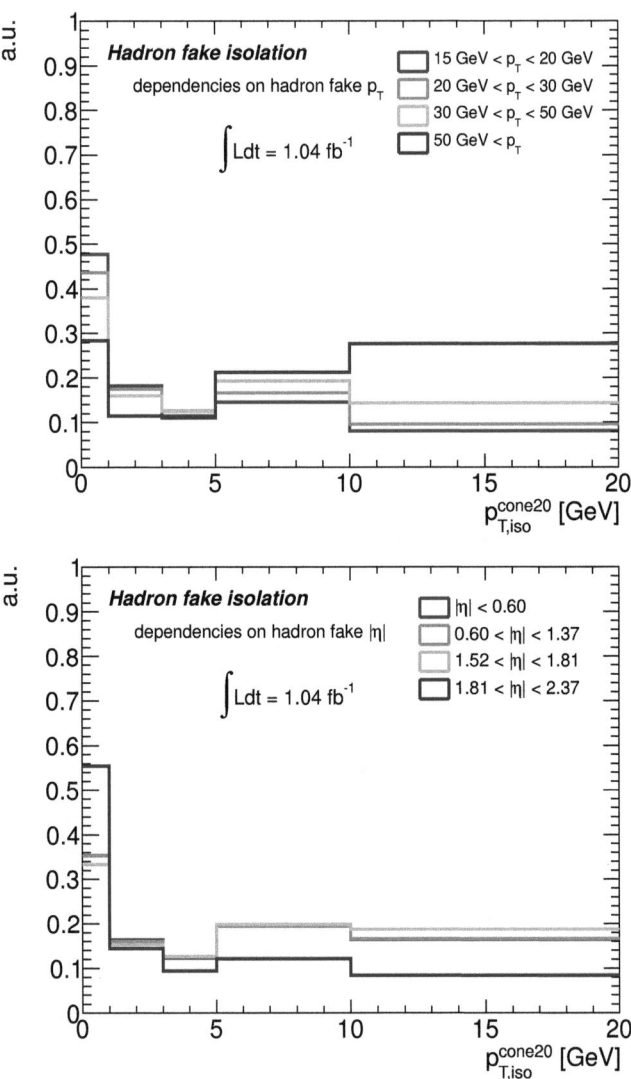

Figure 55: Isolation spectrum of hadron fakes for four bins in p_T (upper plot) and four bins in $|\eta|$ (lower plot). Deviations among all p_T bins can be seen. The isolation shape for $|\eta| > 1.81$ differs w.r.t. the other $|\eta|$ bins.

In order to account for the dependencies of the isolation shapes of hadron fakes on p_T and $|\eta|$ correctly, the final hadron fake template is built from the weighted sum of the templates generated in the four p_T and $|\eta|$ bins.

The photon definition of the $t\bar{t}\gamma$ event selection is replaced by that for hadron fakes and applied on data in order to estimate the expected p_T spectrum of hadron fakes. This fake selection yields 17 events in the e+jets channel and 26 events in the μ+jets channel.

The selected hadron fakes from the e+jets and the μ+jets channel are merged and an exponential function $f_w(p_T)$ is fitted to the p_T spectrum (see Fig. 56). The fit is taken as estimator for the actual p_T spectrum. The contributions w_i of the isolation templates from p_T bin i are calculated by

$$w_i = \frac{\int_{b^i_{\text{low}}}^{b^i_{\text{up}}} f_w(p_T) dp_T}{\int_0^\infty f_w(p_T) dp_T}$$

where b^i_{low} (b^i_{up}) is the lower (upper) limit of bin i.

The isolation templates from the four $|\eta|$ bins are weighted according to the fraction of events selected within $0.0 \leq |\eta| < 1.81$ and $|\eta| > 1.81$ respectively (see right plot in Fig. 56). Hence, the weights for $|\eta|$ bins $1\ldots3$ are all equal ($w_{1\ldots3} = 0.86/3$) while the weight for $|\eta| > 1.81$ reads $w_4 = 0.14$. The final photon isolation templates for prompt photons and for hadron fakes are depicted in Fig. 57.

11.4 Estimation of Prompt Photon Background Contributions

This section describes the methods to determine the background sources of prompt photons. The contributions from electron misidentified as photons, W+jets and multijet (QCD) backgrounds are estimated from data, all other contributions have to be extracted from MC simulation.

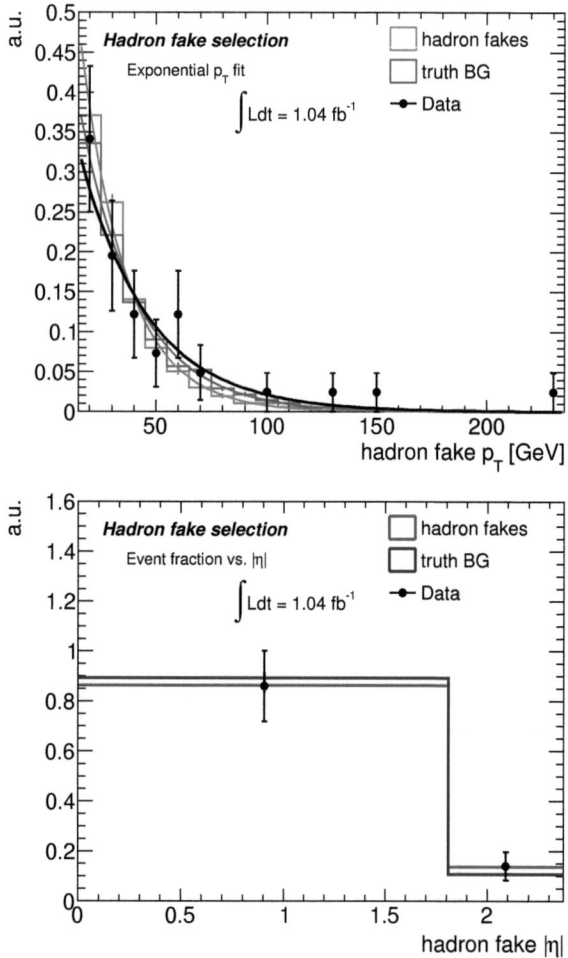

Figure 56: Distribution of transverse momenta (upper plot) and pseudo-rapidities (lower plot) of hadron fakes. The comparison between data and the MC@NLO $t\bar{t}$ sample is also shown, once using the reconstructed hadron fakes and once using the truth information if the photon had been radiated from hadronic decays. Data and MC expectations are in good agreement; the congruence of the shapes of truth hadron fakes and reconstructed hadron fakes indicates that the hadron fake definition established in Sec. 11.1 is suitable.

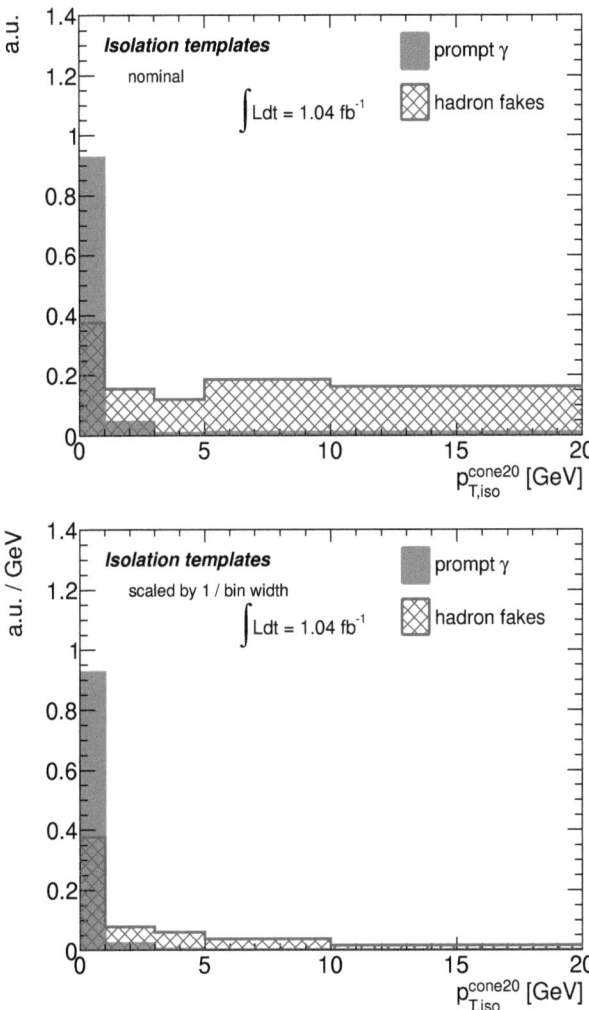

Figure 57: Final prompt photon and hadron fake isolation templates after the application of the MC correction on the prompt photon template and the $|\eta|$ and p_T weighting of the hadron fake template. The upper plot shows the nominal templates, normalized to unity. The entries of the histograms in the lower plot have additionally been scaled by their corresponding bin widths. The last bin of the templates contains overflows.

11.4.1 Background from Electrons Mis-identified as Photons

Since electrons and photons start with the same reconstruction algorithm by finding clusters in the LAr EM calorimeter, electrons might be mis-identified as photons when their tracks could only be reconstructed poorly or not be found by the reconstruction algorithm. Besides, physical processes such as bremsstrahlung and photon conversion can cause difficulties to differentiate between the two objects [141]. The process $Z \to e^+e^-$ is a reliable indicator for electrons faking photons. Within the Z boson mass window, the invariant mass calculated from two e/γ objects is dominated by $Z \to e^+e^-$ decays. Therefore, detecting a photon that yields the Z boson mass together with an electron can considered to be an electron mis-identified as a photon ($e \to \gamma$ fake).

Principle of the Determination of the Photon Fake Rate The number of reconstructed electron pairs in the Z boson mass window is given by

$$N(ee) = \epsilon_1 \epsilon_2 N(ee)_{\text{true}} \qquad (143)$$

where $\epsilon_{1(2)}$ is the efficiency of the first(second) electron to be reconstructed. Similarly, the number of true ee events being reconstructed as an $e\gamma$ event is

$$N(e\gamma) = \epsilon_3 \rho_{e \to \gamma} N(ee)_{\text{true}} \qquad (144)$$

with ϵ_3 being the efficiency of the electron being reconstructed. $\rho_{e \to \gamma}$ is the $e \to \gamma$ fake rate, i.e. the probability that the sub-leading p_T electron is mis-identified as a photon.

From (143) and (144), the $e \to \gamma$ fake rate can be calculated by

$$\rho_{e \to \gamma} = \frac{\epsilon_1 \epsilon_2}{\epsilon_3} \frac{N(e\gamma)}{N(ee)}. \qquad (145)$$

11 CROSS SECTION MEASUREMENT 187

$N(ee)$ and $N(e\gamma)$ can be extracted from data, for the determination of the efficiencies $\epsilon_{1...3}$, a MC based truth matching method is used.

Event Selection The electron trigger is required and the corresponding trigger object has to match the leading p_T electron in order to suppress QCD di-jet events on the one hand and keeping the sub-leading p_T electron unbiased on the other hand. Secondly, $Z \to ee$ events consisting of two reconstructed electrons should be as well selected as $Z \to e\gamma$ events where there might exist only the triggered electron and one photon.

Events after the application of this event selection are most likely originating from $Z \to ee$ decays, either with the Z boson mass reconstructed from two electrons or from one electron and one photon. The event selection for the $e \to \gamma$ fake rate measurement considers both cases:

- The electron trigger had been fired.

- The leading p_T electron has to be tagged as tight electron and match the corresponding trigger object. A transverse momentum p_T of $\geq 25\,\text{GeV}$ is required in order to select only electrons from the trigger plateau region.

- for selecting both $Z \to ee$ <u>and</u> $Z \to e\gamma$ events:

  – There has to be a sub-leading p_T electron with $p_T \geq 15\,\text{GeV}$ which is tagged to be tight or

  – there has to be at least one good photon with $p_T \geq 15\,\text{GeV}$.

- Events where one electrons or photon had been reconstructed within the problematic calorimeter region (LAr hardware failure, see Sec. 9.8) are removed.

Electron Efficiencies The electron efficiencies needed in Eq. (145) are calculated in a $\eta \times p_\mathrm{T}$ matrix, considering 16 bins in pseudorapidity. Since the leading electron has a higher p_T threshold than the sub-leading one, the leading electron efficiency ($\epsilon_1(\eta, p_\mathrm{T})$) is determined in three p_T bins ($25\,\mathrm{GeV} \leq p_\mathrm{T} < 30\,\mathrm{GeV}$, $30\,\mathrm{GeV} \leq p_\mathrm{T} < 50\,\mathrm{GeV}$, $p_\mathrm{T} \geq 50\,\mathrm{GeV}$), whereas the sub-leading electron efficiency $\epsilon_2(\eta, p_\mathrm{T})$ is measured in four p_T bins ($15\,\mathrm{GeV} \leq p_\mathrm{T} < 20\,\mathrm{GeV}$, $20\,\mathrm{GeV} \leq p_\mathrm{T} < 30\,\mathrm{GeV}$, $30\,\mathrm{GeV} \leq p_\mathrm{T} < 50\,\mathrm{GeV}$, $p_\mathrm{T} \geq 50\,\mathrm{GeV}$). A differentiation between ϵ_1 and ϵ_2 has to be made also due the fact that the leading electron is biased by the trigger requirement and the second one is not. Since the electron efficiencies are made available as a function of η and p_T, a distinction between η_3 and η_1 in (145) is not necessary, hence $\epsilon_1 = \epsilon_3 = \epsilon_1(\eta, p_\mathrm{T})$.

The efficiencies are obtained from the $Z \to ee$ MC samples (see Tab. 38). At least one primary vertex with a minimum of five tracks associated to it is required. The electron trigger has to be fired. The truth phase space of the MC simulation is restricted to the acceptance of the electron definition by requiring $|\eta_\mathrm{truth}(e)| < 2.47$ and excluding $1.37 < |\eta_\mathrm{truth}(e)| < 1.52$ and restricting the truth electron momenta to $p_\mathrm{T}^\mathrm{truth}(e_1) \geq 25\,\mathrm{GeV}$ and $p_\mathrm{T}^\mathrm{truth}(e_2) \geq 15\,\mathrm{GeV}$ respectively.

The two truth electrons from the $Z \to ee$ decay are checked if they are reconstructed within a cone of $\Delta R < 0.15$. The reconstructed electron with the higher transverse momentum has to match the truth electron with the higher momentum, the same holds for the sub-leading p_T electron.

If the truth electron is reconstructed and the reconstructed electron fulfils all object definitions, it is weighted by its corresponding trigger, reconstruction an identification efficiency scale factors ($\mathrm{SF}_\mathrm{trig}$, $\mathrm{SF}_\mathrm{reco}$, SF_ID). The sub-leading electron is not biased by the trigger efficiency

11 CROSS SECTION MEASUREMENT

and is therefore only weighted by its corresponding reconstruction and identification scale factors.

Fig. 58 shows the result of the electron efficiency measurement for the leading p_T and the sub-leading p_T electron.

Fake Rate Estimation from MC Simulation The amount of $e \to \gamma$ fakes entering the $t\bar{t}\gamma$ cross section measurement originates mainly from di-leptonic $t\bar{t}$ decays (see Sec. 11.4.4). This number is obtained from the truth electron information of the $t\bar{t}$ MC sample and has to be scaled to the corresponding expectation for data. In order to provide fake rate SFs, a fake rate estimation has to be performed both in $Z \to ee$ MC simulation and in data. The fake rate is evaluated for an invariant mass window of $70 \leq M_{\text{inv}} \leq 120\,\text{GeV}$. The overall MC fake rate is defined as

$$\rho_{e \to \gamma}^{\text{MC}} = \frac{N_{\text{MC}}(e\gamma)}{N_{\text{MC}}^{\text{truth}}(ee)}$$

where $N_{\text{MC}}^{\text{truth}}(ee)$ is the number of all MC events fulfilling the truth phase space requirements. The sub-leading truth electron is tested on reconstructed, selected photons: If the photon is situated within a cone of $\Delta R < 0.15$ around the truth electron, the event is considered for the calculation of $N_{\text{MC}}(e\gamma)$.

The overall $e \to \gamma$ fake rate estimated from MC simulation yields

$$\rho_{e \to \gamma}^{\text{MC}} = (6.70 \pm 0.02)\,\%.$$

Fake Rate Estimation from Data For the $e \to \gamma$ fake rate estimation from data, two invariant mass spectra between $70\,\text{GeV}$ and $120\,\text{GeV}$ are created: One from the leading and sub-leading tight electron and the other one from the leading electron and the leading photon, if that photon has been identified as a good photon. The number

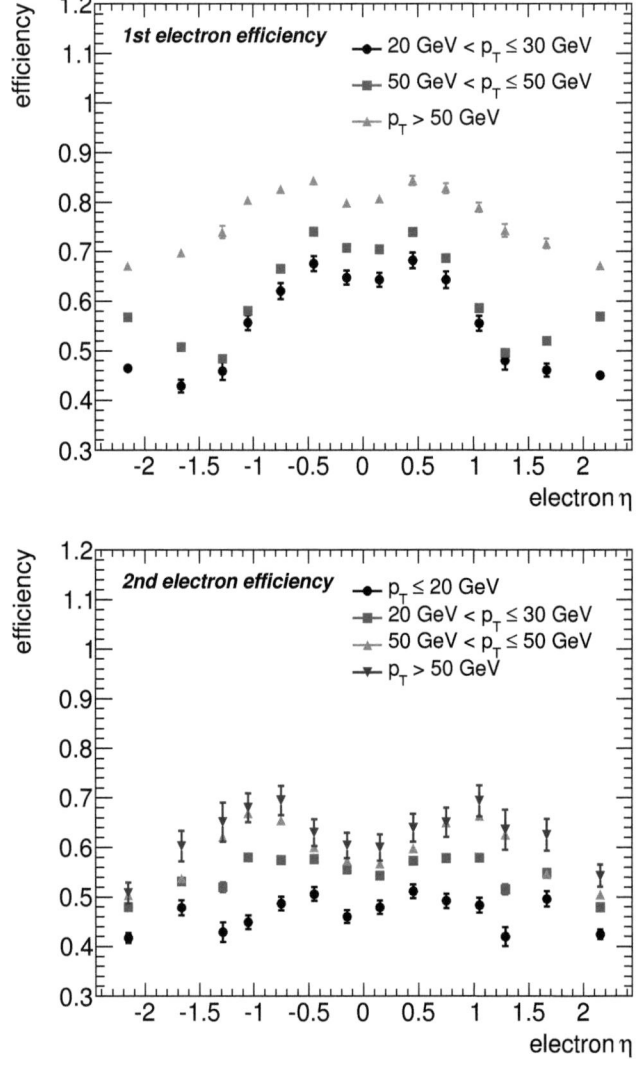

Figure 58: Electron efficiencies obtained from MC simulation for the leading p_T (upper plot) and the sub-leading p_T electron (lower plot).

of $Z \to ee$ events ($N(ee)$) is estimated from the first spectrum, the number of $Z \to e\gamma_{\text{fake}}$ events ($N(e\gamma_{\text{fake}})$) from the second one.

A combined fit is performed on the two invariant mass spectra in order to obtain the overall number of events in the mass window together with the amount of background, which mainly consists of QCD dijet events. The combined fit $f_{\text{comb.}}$ is defined as the sum of a signal function $f_{\text{Sig.}}$ and an exponential background function f_{BG}. The signal function is the convolution of a Crystal Ball (CB) and a Voigtian distribution (VG)[19] ($f_{\text{Sig.}} = f_{\text{CB}} \otimes f_{\text{VG}}$). The combined fit has 9 free parameters: 2 from f_{BG}, 2 from f_{BW}, 4 from f_{CB} and 1 normalization factor for $f_{\text{Sig.}}$. After fitting to data, $N(ee)$ and $N(e\gamma_{\text{fake}})$ are obtained by calculating the integral $\int f_{\text{Sig.}}(M_{\text{inv}}) \, dM_{\text{inv}}$. The integral error is taken from the combined fit in order to take into account the full covariance matrix for the fit uncertainty.

Since the electron efficiencies $\epsilon_{1...3}$ are strongly depending on electron η and p_T (see Fig. 58), the factorization of the fake rate prescription (145) using constant efficiencies is not applicable directly. For that reason, the invariant mass histograms are filled by weighting each entry with the corresponding inverse efficiency:

$$N_{\text{true}}(ee) = \frac{1}{\epsilon_1 \epsilon_2} \sum_i n_i(M_{\text{inv}}, \eta_{1,2}, p_{T,1,2})$$

$$\to N'_{\text{true}}(ee) = \sum_i \frac{n_i(M_{\text{inv}}, \eta_{1,2}, p_{T,1,2})}{\epsilon_1(\eta_1, p_{T,1}) \epsilon_2(\eta_2, p_{T,2})},$$

$$N_{\text{true}}(e\gamma_{\text{fake}}) = \frac{1}{\epsilon_3} \sum_i n_i(M_{\text{inv}}, \eta_1, p_{T,1})$$

$$\to N'_{\text{true}}(e\gamma) = \sum_i \frac{n_i(M_{\text{inv}}, \eta_1, p_{T,1})}{\epsilon_1(\eta_1, p_{T,1})}.$$

---

[19] A Voigtian distribution is the convolution of a Gaussian and a Breit-Wigner distribution.

With this event weighting, the overall $e \to \gamma$ fake rate $\rho_{e\to\gamma}^{\text{Data}}$ directly reads

$$\rho_{e\to\gamma}^{\text{Data}} = \frac{N'_{\text{true}}(e\gamma_{\text{fake}})}{N'_{\text{true}}(ee)}. \tag{146}$$

The integrals of the fits yield $N'(ee) = 903{,}000 \pm 2{,}000$ and $N'(e\gamma) = 48{,}700 \pm 500$ respectively. The result for the estimated $e \to \gamma$ fake rate from data is

$$\rho_{\text{Data}} = (5.40 \pm 0.06)\,\%.$$

There is a large discrepancy between the results for the overall fake rates obtained from data and MC. One reason is that it is not possible to restrict the true phase space of the MC $Z \to ee$ sample in such a way that it matches the detector acceptance perfectly. Secondly, the truth matching method for obtaining the MC fake rate completely neglects the possibility that also jet→electron fakes from $Z \to j_1 j_2$ are measured in data.

The origins of the considered systematic uncertainties and the methods of their determination are described in the next section. Fig. 59 shows the weighted invariant mass spectra and the results of the combined fits.

Dependency on Photon Pseudo-Rapidity and Momentum In order to improve the precision of the correction of the $e \to \gamma$ fake rate from MC simulation to data, the fake rates have been determined in two bins of photon p_T and three bins of photon pseudo-rapidity $|\eta_{S2}|$. The comparison between data and MC prediction for the fake rate is shown in Fig. 60.

Discussion of Systematic Uncertainties Three possible systematic effects have been investigated for the determination of the $e \to \gamma$ fake rate: The impact of the fit uncertainties ($f_{\text{comb.}}$), pile-up effects and uncertainties from the estimation of the electron efficiencies. The fit uncertainties of $f_{\text{comb.}}$ are in the range from $1.3\,\%$ to $4.3\,\%$.

11 CROSS SECTION MEASUREMENT

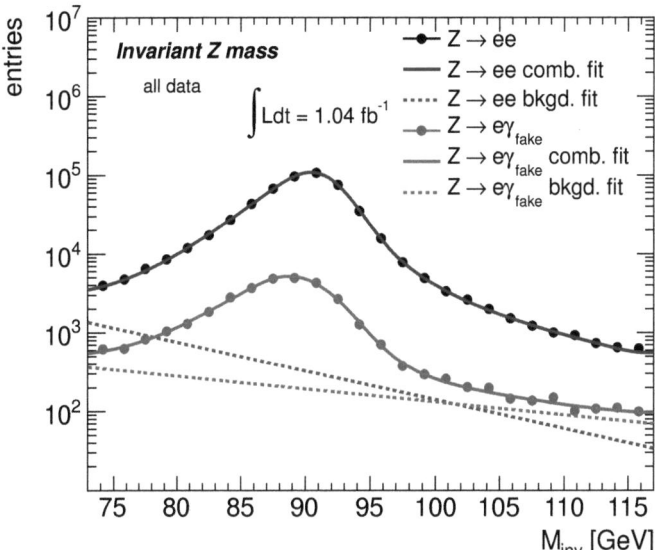

Figure 59: Invariant mass distributions of ee and $e\gamma_{\text{fake}}$ events from data together with their combined fits ($f_{\text{comb.}}$). The exponential background of the fits is depicted by the dashed lines.

The impact of pileup on the fake rate is studied by comparing three invariant mass spectra for $N'(ee)$ and $N'(e\gamma)$ in three bins of number of primary N_{PV} vertices ($0 \leq N_{\text{PV}} < 3$, $3 \leq N_{\text{PV}} \leq 4$, $N_{\text{PV}} \geq 5$). The fake rates ρ^i_{pileup} ($i = 1\ldots 3$) are obtained again from the integrals of combined fits $f_{\text{comb.}}$. A constant fit is performed on the ρ^i_{pileup} to obtain the weighted average.

Fig. 61 shows the result for the ρ^i_{pileup}, indicating that the uncertainties due to pile-up effects are in the same order as the statistical uncertainties of $\rho^{\text{Data}}_{e \to \gamma}$. Furthermore, no systematic behavior depending on the number of primary vertices can be observed. Hence, an additional uncertainty due to pile-up effects is not included.

For the determination of the uncertainties of the electron efficiencies $\epsilon_{1\ldots 3}$, the electron efficiencies have been recalculated by varying the re-

11.4 Estimation of Prompt Photon Background Contributions

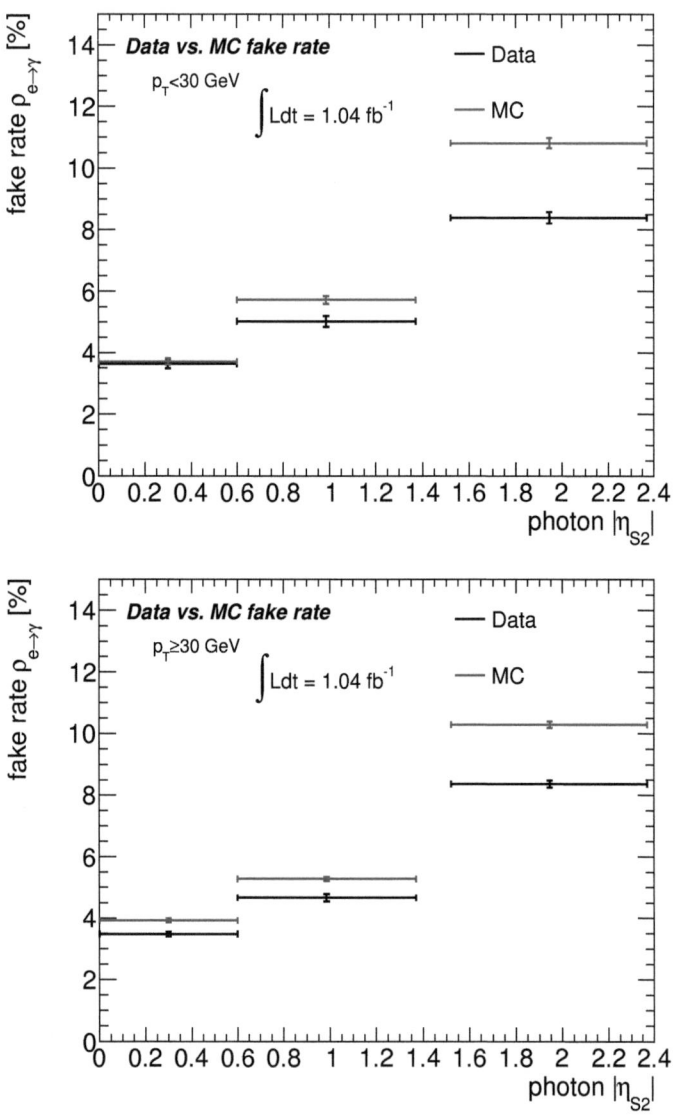

Figure 60: Comparison of the $e \to \gamma$ fake rate in 2×3 bins in photon p_T and $|\eta_{S2}|$.

construction, trigger and ID scale factors up and down by one standard deviation of their corresponding uncertainties. This alters the electron efficiencies in each bin in p_T and η independently. The weighted average of the relative deviations in all bins has been evaluated, weighting the summands with the corresponding relative bin content. This yields a relative uncertainty of 3.5 % for the leading p_T electron efficiency and 2.8 % for the sub-leading electron. Regarding $\frac{\epsilon_1 \epsilon_2}{\epsilon_3}$ in Eq. (145) and assuming $\epsilon_1 = \epsilon_3$, the overall uncertainty of the electron efficiencies reads 5.7 %.

The statistical uncertainties of the electron efficiencies are considered to be a systematic effect either. The overall statistical uncertainty has been determined the same way as the systematic uncertainties of the scale factors and read 1.5 % for the leading and 2.5 % for the sub-leading electron. The overall statistical uncertainty is 3.3 %.

The statistical uncertainties due to the combined fit to the invariant mass spectra are already included in Fig. 60. All considered uncertainties are broken down in Tab. 10.

Description	rel. uncertainty
electron trigger, ID and reconstruction SFs	5.7 %
stat. uncertainty of electron efficiency measurement	3.3 %
fit uncertainty	1.3 % - 4.3 %

Table 10: Breakdown of systematic uncertainties taken into account for the $e \to \gamma$ fake rate scale factors.

Derivation of Fake Rate Scale Factors As already mentioned in the beginning of the section, the expected yield of electrons misidentified as photons in MC simulations for various processes has to be extrapolated to the corresponding expectation in data. With the

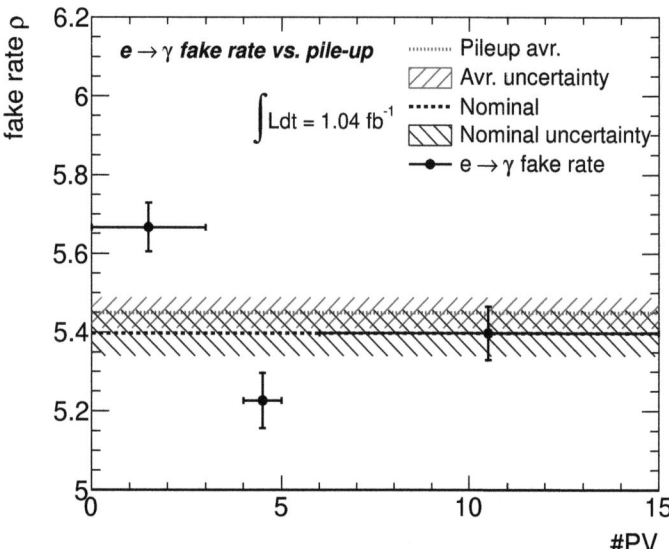

Figure 61: Result for the $e \to \gamma$ fake rate measured in three bins of number of primary vertices. The black dashed line depicts the nominal fake rate $\rho_{e \to \gamma}^{\text{Data}}$, the red dotted line the average fake rate of the three bins of number of primary vertices. The red shaded area represents the fit uncertainty of the pile-up average, the other dashed area depicts the fit uncertainty of the overall $\rho_{e \to \gamma}^{\text{Data}}$.

results shown in Fig. 60, $e \to \gamma$ fake rate scale factors can be derived by

$$\text{SF}_{e \to \gamma_{\text{fake}}}(p_{\text{T}}, |\eta|) = \frac{\rho_{e \to \gamma}^{\text{Data}}(p_{\text{T}}, |\eta|)}{\rho_{e \to \gamma}^{\text{MC}}(p_{\text{T}}, |\eta|)}. \quad (147)$$

These scale factors will be applied later on events in the MC samples where an electron is produced in the hard process but is reconstructed as a photon. The scale factors derived from the $e \to \gamma$ fake rate measurement in two bins in photon p_{T} and three bins in photon $|\eta_{\text{S2}}|$ are shown in Fig. 62.

11 CROSS SECTION MEASUREMENT

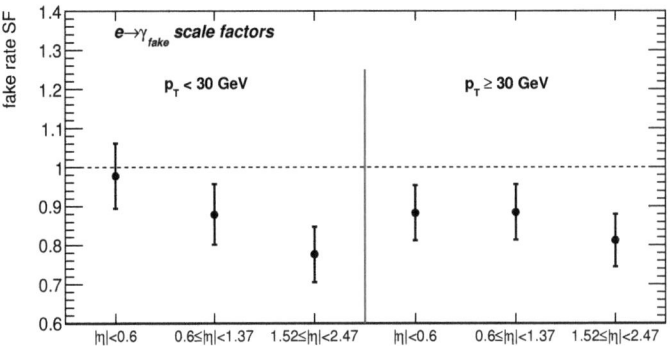

Figure 62: Scale factors derived from the $e \to \gamma$ fake rate measurement, split in two bins in photon p_T and three bins in photon $|\eta_{S2}|$. The error bars include all uncertainties. The horizontal, dashed line marks the equality of ρ_{Data} and ρ_{MC}.

11.4.2 W+Jets+γ Background

The estimation of the W+jets contribution to the prompt photon background is estimated from data. For this purpose, a control region (CR) orthogonal to the $t\bar{t}\gamma$ event selection is defined. The W+jets+γ selection is the same as the standard event selection described in Sec. 9.9, but requires at least two jets but less than four jets with $p_T > 25\,\text{GeV}$ and $|\eta| < 2.5$ instead. Additionally, a veto on the existence of b-tagged jets is applied in order to reduce contributions from top quark production. The sample obtained from this modified event selection is enriched with events from W+jets production. Remaining contributions of $t\bar{t}$ and $t\bar{t}\gamma$ events in the CR are estimated from MC simulation, the number of selected events in the CR is given in Tab. 11.

The yield of W+jets events that exhibit a photon is extrapolated to the expectation in the signal region (SR) using dedicated W+jets+γ

	e+jets channel	μ+jets channel
$N^{\text{CR}}_{W+\text{jets}+\gamma}$	1252 ± 35	2174 ± 47
1-jet bin	813 ± 29	1551 ± 40
2-jet bin	318 ± 18	488 ± 22
3-jet bin	121 ± 12	135 ± 12
$t\bar{t}$ leakage	20.5 ± 0.4 ± 2.4	40.5 ± 0.6 ± 3.7
$t\bar{t}\gamma$ leakage	8.1 ± 0.2 ± 0.8	14.5 ± 0.2 ± 1.4
$N^{\text{exp,CR}}_{W+\text{jets}+\gamma}$	1223 ± 35 ± 3	2119 ± 47 ± 4
$N^{\text{CR}}_{W+\text{jets}+\gamma,\text{MC}}$ [10^{-2}]	52.2 ± 0.7 ± 8.8	130 ± 1 ± 21
$N^{\text{SR}}_{W+\text{jets}+\gamma,\text{MC}}$ [10^{-2}]	0.21 ± 0.03 ± 0.05	0.35 ± 0.04 ± 0.08
r_{MC} [10^{-3}]	4.1 ± 0.5 ± 1.2	2.7 ± 0.3 ± 0.8
f_γ	0.70 ± 0.06	0.73 ± 0.05
$N^{\text{exp,SR}}_{W+\text{jets}+\gamma}$	**3.5 ± 0.6 ± 1.0**	**4.2 ± 0.5 ± 1.2**

Table 11: Event yields of the W+jets+γ background estimation and extrapolation factor r_{MC} for the e+jets and μ+jets channel. The uncertainty of the final expectation $N^{\text{exp,SR}}_{W+\text{jets}+\gamma}$ is dominated by the uncertainty of r_{MC}, originating from Berends-Giele scaling. The overall number $N^{\text{CR}}_{W+\text{jets}+\gamma}$ obtained for all jet multiplicities is additionally split into different jet bins (cf. Fig. 63).

MC samples by multiplying the data yield in the CR with the ratio r_{MC} of expected W+jets+γ yield in the SR over the yield in the CR:

$$r_{\text{MC}} = \frac{N^{\text{SR}}_{W+\text{jets}+\gamma,\text{MC}}}{N^{\text{CR}}_{W+\text{jets}+\gamma,\text{MC}}}. \qquad (148)$$

Hence, the expected amount of W+jets+γ in the SR is then given by

$$N^{\text{SR}}_{W+\text{jets}+\gamma} = N^{\text{CR}}_{W+\text{jets}+\gamma} \cdot r_{\text{MC}}. \qquad (149)$$

The results for the event yields in the CR and r_{MC} for the e+jets and μ+jets channel are broken up in Tab. 11.

$N^{\text{SR}}_{W+\text{jets}+\gamma}$ still contains both prompt photon and hadron fake candidates. Since hadron fakes are already fully covered by the hadron fake template (Sec. 11.3.3), the amount of prompt photons in $N^{\text{SR}}_{W+\text{jets}+\gamma}$ is calculated by performing a template fit on the isolation distribution

11 CROSS SECTION MEASUREMENT

of selected W+jets+γ from data in the CR. The fraction f_γ of prompt photons estimated from the template fit is additionally multiplied on $N^{\text{SR}}_{W+\text{jets}+\gamma}$ in order to obtain the fraction of W+jets+γ events in the SR that contain only prompt photons.

Since the event selection in the CR considers three or less jets, whereas the event selection in the SR requires at least four jets in contrast, the dependence of f_γ on the jet multiplicity has been investigated by performing the template fit for events containing exactly one, two and three jets in the CR. Fig. 63 indicates that there is no dependence on the jet multiplicity, hence f_γ is assumed to be able to be used to estimate the fraction of prompt photons in the SR (≥ 4 jets). Fig. 64 shows the result of the template fit for obtaining f_γ in the CR for all jet multiplicities together.

The final results for the data-driven W+jets+γ background estimation are given in Tab. 11. The main source of uncertainties affecting the W+jets+γ background estimation is the Berends-Giele scaling which translates to the estimation of the scale factor r_{MC} of expected W+jets+γ events from the CR to the SR. Other uncertainties considered for r_{MC} are the uncertainties of the $t\bar{t}$ and $t\bar{t}\gamma$ simulation as described in Sec. 10.7 and 12.3.7; the uncertainty on the $t\bar{t}\gamma$ additionally includes the uncertainty on the k-factor as described in Sec. 12.3.8. The uncertainties of the estimation of the leakage of events from other processes than W+jets+γ into the CR have a minor effect, as well as the fit uncertainty of the estimation of the fraction of prompt photons f_γ.

11.4.3 Multijet Background (QCD+γ)

Multijet events may mimic an isolated lepton when a lepton from hadronic decays is radiated outside the jet cone and thus fulfils the isolation criterion. Such events may also exhibit a prompt photon and hence survive the full $t\bar{t}\gamma$ event selection.

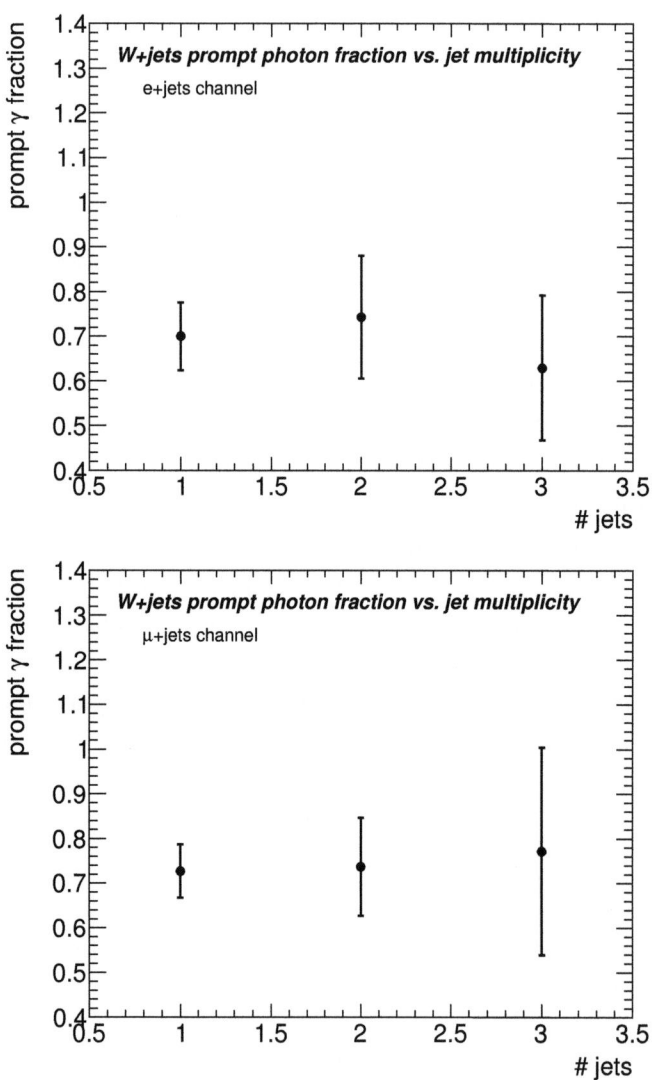

Figure 63: Fraction of prompt photons f_γ obtained from template fits in the CR for events containing one, two and three jets. No dependence on the jet multiplicity can be observed.

11 CROSS SECTION MEASUREMENT

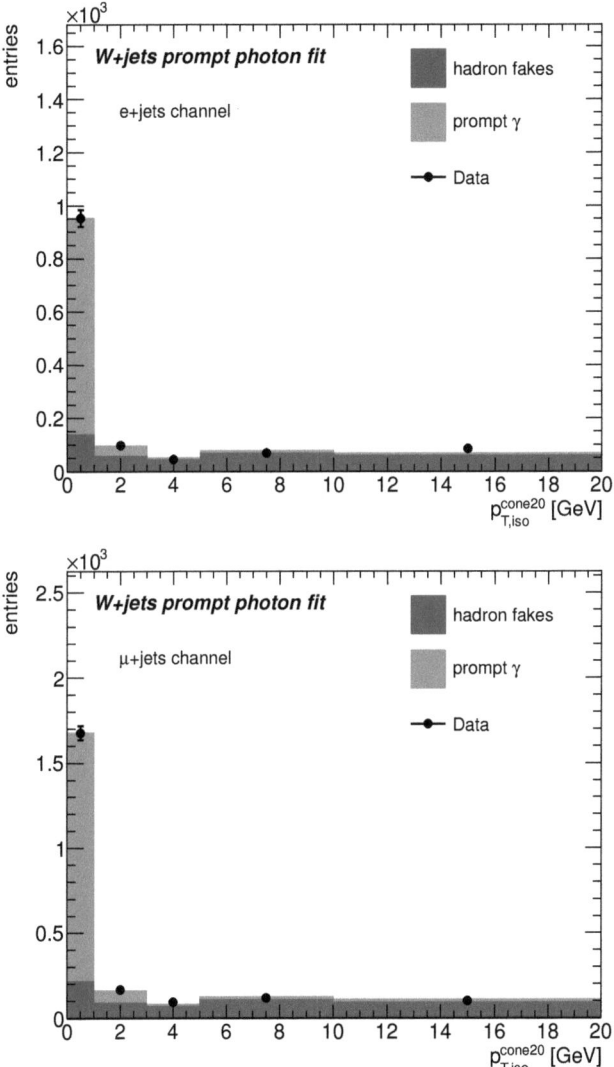

Figure 64: Template fit of W+jets+γ in the CR for obtaining f_γ for the e+jets channel (upper plot) and the μ+jets channel (left plot).

11.4 Estimation of Prompt Photon Background Contributions

The multijet background is derived from a matrix method in the e+jets and in the μ+jets channel. The result of this estimation is provided centrally in the ATLAS top quark physics working group [142]. The amount of multijet events containing (prompt) photons is applied on top of that official multijet estimation.

Matrix Method The matrix method is based on the hypothesis that in a dedicated control region enhanced by multijet events all tight leptons are fakes. The fraction $\epsilon_{\text{fake}} = N_{\text{fake}}^{\text{tight}}/N_{\text{fake}}^{\text{loose}}$ of loose leptons in the CR that also pass the tight lepton criteria is assumed to be the same in the SR. The according efficiency $\epsilon_{\text{real}} = N_{\text{real}}^{\text{tight}}/N_{\text{real}}^{\text{loose}}$ for a real loose lepton to be identified also as a tight lepton is measured from $Z \to \ell^+\ell^-$ decays in data and from W+jets MC simulation.
The overall number of loose leptons in the SR can be written as

$$N^{\text{loose}} = N_{\text{real}}^{\text{loose}} + N_{\text{fake}}^{\text{loose}} \tag{150}$$

and for tight leptons

$$N^{\text{tight}} = N_{\text{real}}^{\text{tight}} + N_{\text{fake}}^{\text{tight}} = \epsilon_{\text{real}} N_{\text{real}}^{\text{loose}} + \epsilon_{\text{fake}} N_{\text{fake}}^{\text{loose}} \tag{151}$$

respectively. With these definitions, the amount of multijet events in the SR that exhibit a tight lepton can be written as

$$N_{\text{fake}}^{\text{tight}} = \frac{\epsilon_{\text{fake}}}{\epsilon_{\text{real}} - \epsilon_{\text{fake}}}(N_{\text{loose}}\epsilon_{\text{real}} - N_{\text{tight}}). \tag{152}$$

N_{loose} and N_{tight} are estimated from data in the SR by applying once the standard (tight) lepton definition and once the loose lepton definition instead of the standard definition.
The loose electron is defined as a medium electron (see Sec. 7.1.1) with a loosened restriction of $E_T^{\text{cone20}} < 6\,\text{GeV}$ (instead of $E_T^{\text{cone20}} < 3.5\,\text{GeV}$) on the calorimeter isolation. Additionally, the standard $\not{E}_T$ definition is replaced by that considering medium electrons since this

11 CROSS SECTION MEASUREMENT

definition is also used in the CR to determine ϵ_fake. The medium $\not{E}_\text{T}$ in the loose selection is required to be $5\,\text{GeV} < \not{E}_\text{T} < 20\,\text{GeV}$.
For the μ+jets channel, the loose muon is defined by removing the p_T^cone30 and E_T^cone30 isolation requirements. The $\not{E}_\text{T}$ definition is not modified. The requirement of $\not{E}_\text{T} > 20\,\text{GeV}$ is removed and replaced by an inverted requirement on the transverse W mass $M_\text{T}(W) < 20\,\text{GeV}$. The triangular cut $M_\text{T}(W) + \not{E}_\text{T}$ is inverted and required to be $M_\text{T}(W) + \not{E}_\text{T} < 60\,\text{GeV}$.

QCD+γ Estimation In order to estimate the fraction f_γ^all of multijet events exhibiting an additional photon w.r.t. to multijet events without requiring a photon, the event selection in the CR is performed once with and once without the requirement having at least one selected photon. The fraction of number of events with over the number of events without a photon $f_\gamma^\text{all} = N_{\text{QCD}+\gamma}/N_\text{QCD}$ is then extrapolated to the SR.
The leakage of events from processes other than QCD multijet production is estimated from MC simulations and subtracted from N_QCD and $N_{\text{QCD}+\gamma}$ respectively. The leakage has been estimated for $t\bar{t}$, $t\bar{t}\gamma$, W+jets and Z+jets samples. The several contributions of leakages are applied the uncertainties described in Sec. 10.7; in particular, the uncertainty due to Berends-Giele scaling for Z+jets and W+jets processes; the expected amount of $t\bar{t}$ and $t\bar{t}\gamma$ events are subject to the uncertainties of MC modeling as described in Sec. 12.3.7. In addition, the amount of $t\bar{t}\gamma$ leakage is applied the uncertainty of the k-factor (Sec. 12.3.8).
All relevant numbers for the calculation of f_γ^all are broken down in Tab. 12.
A template fit to the selected multijet events with an additional photon in the CR is performed in order to find the fraction f_γ^prompt of prompt photons. Hence, the final fraction of prompt photons f_γ in the CR is given by $f_\gamma = f_\gamma^\text{all} \cdot f_\gamma^\text{prompt}$.

	e+jets channel		μ+jets channel	
	before photon	after photon	before photon	after photon
$N^{\text{all}}_{\text{QCD}(+\gamma)}$	62300 ± 250	172 ± 13	107500 ± 300	109 ± 10
Z+jets leakage	$1390 \pm 200 \pm 210$	$38 \pm 2 \pm 6$	$430 \pm 65 \pm 70$	$10 \pm 1 \pm 2$
W+jets leakage	$1920 \pm 290 \pm 300$	$3 \pm 1 \pm 1$	$550 \pm 85 \pm 90$	$1 \pm 1 \pm 1$
$t\bar{t}$ leakage	$1020 \pm 180 \pm 120$	$9 \pm 1 \pm 1$	$710 \pm 120 \pm 65$	$9 \pm 1 \pm 1$
$t\bar{t}\gamma$ leakage	$20 \pm 3 \pm 3$	$6 \pm 1 \pm 1$	$13 \pm 2 \pm 2$	$4 \pm 1 \pm 1$
$N_{\text{QCD}(+\gamma)}$	$57900 \pm 460 \pm 390$	$116 \pm 22 \pm 6$	$105800 \pm 370 \pm 130$	$84 \pm 19 \pm 2$
	$f^{\text{all}}_\gamma = (2.01 \pm 0.37 \pm 0.11) \cdot 10^{-3}$		$f^{\text{all}}_\gamma = (0.79 \pm 0.18 \pm 0.02) \cdot 10^{-4}$	

Table 12: Estimation of the fraction of prompt photons f^{all}_γ in the multijet control region shown together will all relevant event yields/leakages.

Fig. 65 shows the result of the template fit for obtaining f^{prompt}_γ for the e+jets and μ+jets channel.
The estimated number of multijet events in the SR and the fraction f^{all}_γ estimated from the CR and the prompt photon fraction f^{prompt}_γ are listed in Tab. 13. An uncertainty of 100 % on the multijet expectation (see Sec. 10.7) is applied on $N^{\text{SR}}_{\text{QCD}}$ and yields 280 ± 280 in the e+jets channel and 500 ± 500 in the μ+jets channel. After the multiplication with f^{all}_γ and f^{prompt}_γ, the final expectation $N^{\text{SR}}_{\text{QCD}+\gamma}$ of multijet events with the production of an additional prompt photon yields $0.39^{+0.40}_{-0.39}(\text{stat.}) \pm 0.39(\text{syst.})$ in the e+jets channel and $0.24^{+0.25}_{-0.24}(\text{stat.}) \pm 0.24(\text{syst.})$ in the μ+jets channel.
These are the final numbers for the QCD+γ background contribution that enter the template fit (Sec. 11.5).

	e+jets channel	μ+jets channel
$N^{\text{SR}}_{\text{QCD}}$	280 ± 280	500 ± 500
f^{all}_γ $[10^{-4}]$	$20.1 \pm 3.7 \pm 1.1$	$0.79 \pm 0.18 \pm 0.02$
f^{prompt}_γ	0.70 ± 0.11	0.62 ± 0.12
$N^{\text{SR}}_{\text{QCD}+\gamma}$	$\mathbf{0.39^{+0.40}_{-0.39} \pm 0.39}$	$\mathbf{0.24^{+0.25}_{-0.24} \pm 0.24}$

Table 13: Multijet background yields in the SR for the e+jets and μ+jets channel. Note that an uncertainty of 100 % is applied on $N^{\text{SR}}_{\text{QCD}}$ *before* the multiplication with f^{all}_γ and f^{prompt}_γ.

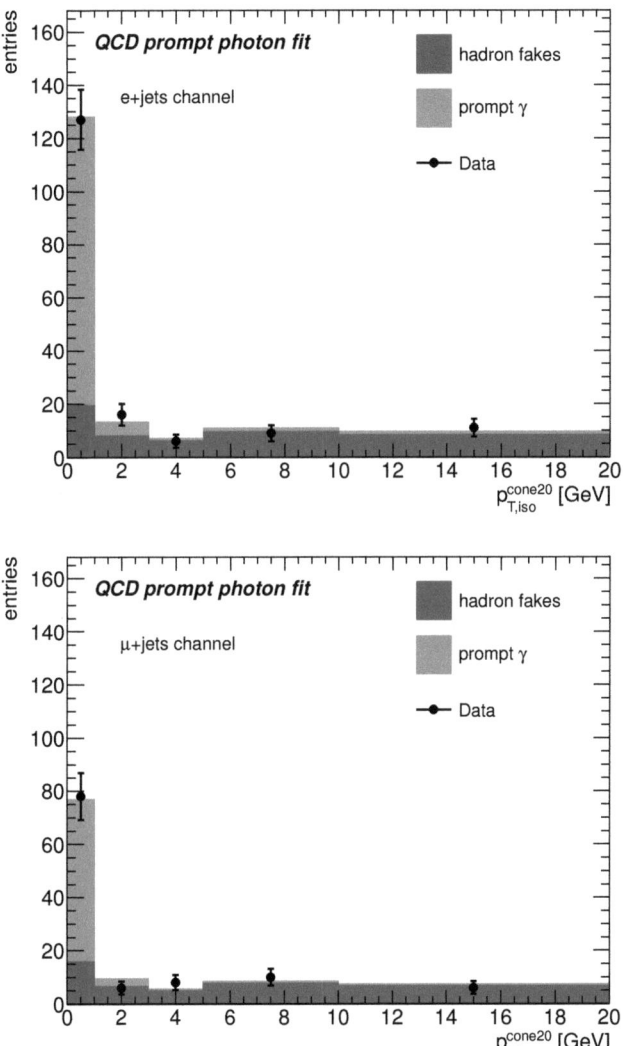

Figure 65: Template fit in the CR for obtaining f_γ^{prompt} for the e+jets channel (upper plot) and the μ+jets channel (lower plot). Note that f_γ^{prompt} is calculated from $N_{\text{QCD}+\gamma}^{\text{all}}$, i.e. before the subtraction of any leakage.

11.4.4 $t\bar{t}$ Background

The outcome of the $t\bar{t}\gamma$ cross section measurement is directly proportional to the inverse of the signal event selection efficiency which is completely determined by the WHIZARD $t\bar{t}\gamma$ MC simulation.

Hence, remaining MC@NLO $t\bar{t}$ events surviving the event selection after having removed contributions that would fall into the signal phase space (see Sec. 10.4) are considered to be background. There are two kinds of background events that have to be considered:

$t\bar{t}$ Events Containing Real Photons Real photons in the remaining MC@NLO $t\bar{t}$ events can have different origins and have to be treated in a different manner.

First, the reconstructed photon is tested if it is matched to a real truth photon within a cone size of $\Delta\eta \times \Delta\phi = 0.025 \times 0.050$ space[20]. Matched truth photons with a transverse momentum of $p_T < 10\,\text{GeV}$ are not considered to be the real origin of the reconstructed photon but are believed to be matched accidentally.

Photons with $p_T \geq 10\,\text{GeV}$ are then further investigated and can either originate from the hard process which includes bremsstrahlung, where a quark emits a photon before/during parton showering. If the photon is found to be radiated within the hard process, the origin of its corresponding truth photon is investigated and always found to be radiated from either a truth lepton or a quark (bremsstrahlung).

Tab. 14 lists all contributions of remaining $t\bar{t}$ photons together with their origins. The first quoted uncertainty is the statistical uncertainty, the second one contains the uncertainties of the MC@NLO $t\bar{t}$ modeling as described in Sec. 12.3.7.

All $t\bar{t}$ events that contain a real prompt photon but are simulated outside the signal phase space defined by the invariant mass cuts of the

---

[20]The photon classification and matching is performed externally by the MCTruthClassifier tool [143].

WHIZARD $t\bar{t}\gamma$ signal MC sample and survive the full event selection, are di-leptonic $t\bar{t}$ events, where one lepton is correctly identified and fulfils the corresponding lepton definitions, whereas the second lepton is not identified but radiates a photon, which fulfils the photon definitions. In the case that the photon is radiated from a close-by electron, the two objects cannot be distinguished experimentally at all, hence such photons are treated as $e \to \gamma$ fakes.

The second contribution is bremsstrahlung, where the photon is radiated from quarks before hadronization. In this case, only semi-leptonic $t\bar{t}$ events can pass the event selection and are considered to be prompt photon contributions either.

The contributions from lepton radiation and bremsstrahlung sum up to the overall number of photons with the requirement of $p_T^{\text{truth}} > 10\,\text{GeV}$. As consequence, the remaining photons are hadron fakes and hence are omitted since hadron fakes are already fully covered by the hadron fake template.

The remaining prompt photons enter the final template fit as prompt photon background source; photons that have their origin from radiation from electrons are applied the corresponding $e \to \gamma_{\text{fake}}$ scale factor (see Sec. 11.4.1).

Electrons Mis-identified as Photons The major background contribution of MC@NLO $t\bar{t}$ events are those, where no real physical photon is produced, but physical electrons are mis-identified and reconstructed as photons ($e \to \gamma_{\text{fake}}$ background, see Sec. 11.4.1). Since $t\bar{t}$ events give rise to the largest selection efficiency w.r.t. the event preselection and the cross section of $t\bar{t}$ production without the additional radiation of a real, prompt photon is ≈ 50 times higher than the expected $t\bar{t}\gamma$ cross section, this kind of background makes up the largest background contribution.

In order to estimate the amount of $t\bar{t}$ events that survive the full event selection due to mis-identified electrons, truth prompt electrons from

Photon source	e+jets channel	μ+jets channel
all events	$2.82 \pm 0.12 \pm 0.33$	$3.94 \pm 0.14 \pm 0.36$
events $(p_\mathrm{T}^\mathrm{truth}(\gamma) > 10\,\mathrm{GeV})$	$0.76 \pm 0.06 \pm 0.09$	$1.44 \pm 0.09 \pm 0.13$
radiation from e	$0.51 \pm 0.05 \pm 0.06$	$0.86 \pm 0.07 \pm 0.08$
radiation from μ	$0.16 \pm 0.02 \pm 0.02$	$0.37 \pm 0.05 \pm 0.03$
radiation from τ	$0.076 \pm 0.024 \pm 0.009$	$0.087 \pm 0.025 \pm 0.008$
di-leptonic events	$0.75 \pm 0.06 \pm 0.09$	$1.32 \pm 0.09 \pm 0.12$
bremsstrahlung	$0.015 \pm 0.005 \pm 0.002$	$0.13 \pm 0.04 \pm 0.01$

Table 14: Contributions of remaining MC@NLO $t\bar{t}$ events after the $t\bar{t}\gamma$ signal phase space overlap removal, i.e. contributions that contain real prompt photons but are not covered by the signal phase space definition. The numbers are shown for the e+jets and μ+jets channel. The first quoted uncertainty is the statistical uncertainty, the second one represents the systematic uncertainty of the MC@NLO $t\bar{t}$ modeling as described in Sec. 12.3.7. The statistical uncertainties are taken into account for the Gaussian prior of the template fit (see Sec. 11.5). Note that in this table, the contribution of photons radiated from electrons has not yet been applied the corresponding $e \to \gamma_\mathrm{fake}$ SF (cf. Tab. 15).

top quark decays $(t \to Wb \to e\nu_e b)$ are inspected for surrounding, reconstructed photons that fulfil the good photon definition established in Sec. 9.4. If the angular distance between the truth electron and the photon is $\Delta R < 0.1$, the event is applied the according $e \to \gamma_\mathrm{fake}$ scale factor (see Eq. (147)).

The contributions of electrons mis-identified as photons in $t\bar{t}$ events are listed in Tab. 15 separately for the several top quark pair decay modes. Mostly di-leptonic $t\bar{t}$ events give rise to the $e \to \gamma_\mathrm{fake}$ background.

11.4.5 Other Background Contributions Estimated from MC Simulation

The remaining non-$t\bar{t}$ background contributions besides W+jets+γ and QCD+γ events originate from Z+jets, di-boson and single top events. These contributions are estimated completely from MC simulation since there is no way to determine them from data. The ex-

11 CROSS SECTION MEASUREMENT

Decay mode	e+jets channel	μ+jets channel
semi-leptonic (e/μ+jets)	$0.040 \pm 0.011 \pm 0.004$	$0.030 \pm 0.013 \pm 0.003$
di-leptonic (e/μ)	$6.3 \pm 0.8 \pm 0.5$	$8.8 \pm 1.1 \pm 0.7$
di-leptonic ($e, \mu + \tau$)	$0.38 \pm 0.06 \pm 0.03$	$0.61 \pm 0.09 \pm 0.05$
Sum	$6.7 \pm 0.8 \pm 0.5$	$9.4 \pm 1.1 \pm 0.7$
radiation from e	$0.45 \pm 0.04 \pm 0.05$	$0.76 \pm 0.06 \pm 0.07$
Overall sum	$7.17 \pm 0.8 \pm 0.5$	$10.2 \pm 1.1 \pm 0.7$

Table 15: Contributions of MC@NLO $t\bar{t}$ events where a truth electron has been misidentified as a photon, shown for the e+jets and μ+jets channel. Prompt photons radiated from electrons, which are treated as $e \to \gamma$ fakes, have been multiplied with the corresponding $e \to \gamma_{\text{fake}}$ SF (cf. Tab. 14). The first uncertainty quoted is the statistical one, the second uncertainty represents the systematic uncertainty of the $e \to \gamma$ fake rate estimation (see Sec. 11.4.1). The statistical uncertainties are taken into account for the Gaussian prior of the template fit (see Sec. 11.5).

pected yields of the remaining non-$t\bar{t}$ background are listed in Tab. 16.

The first quoted uncertainty is the statistical one, the second uncertainty represents the systematic uncertainty due to MC modeling; the MC uncertainties of the Z+jets+γ contribution originate from Berends-Giele scaling, the uncertainties of the di-boson and single top quark yields have been evaluated from theoretical calculations (see Sec. 10.7). The statistical uncertainties are often dominant since only a few events survive the full $t\bar{t}\gamma$ event selection. The reason is that the standard Z+jets MC samples are used instead of a dedicated set of Z+jets+γ MC samples. The latter ones were not available when this thesis was created.

Since the radiation of prompt photons as well as hadron fakes from jet fragmentation is completely based on MC modeling (parton showering and hadronization software), a template fit to derive the fraction of prompt photons cannot not be applied. Prompt photons and hadron fakes are extracted by using MC truth information instead: photons with their parent particles identified as hadrons are classified as hadron fakes whereas photons radiated from leptons, quarks or W/Z bosons

	Isolated photons		$e \to \gamma$ fakes	
	e channel	μ channel	e channel	μ channel
Z+jets+γ	$1.4 \pm 1.1 \pm 0.8$	$1.5 \pm 1.1 \pm 0.6$	$1.0 \pm 0.6 \pm 0.6$	$0 \pm 0 \pm 0$
Di-boson	$0.15^{+0.16}_{-0.15} \pm 0.01$	$0.033 \pm 0.019 \pm 0.002$	$0.022 \pm 0.013 \pm 0.002$	$0 \pm 0 \pm 0$
Single top	$0.59 \pm 0.18 \pm 0.04$	$0.26 \pm 0.13 \pm 0.02$	$0.123 \pm 0.071 \pm 0.008$	$0.052 \pm 0.037 \pm 0.004$
Sum	**$2.2 \pm 1.1 \pm 0.8$**	**$1.8 \pm 1.1 \pm 0.6$**	**$1.1 \pm 0.6 \pm 0.6$**	**$0.052 \pm 0.037 \pm 0.004$**

Table 16: Z+jets+γ, di-boson and single top background contributions estimated from MC simulation. The first quoted uncertainty is the statistical one, the second uncertainty represents the systematic uncertainty due to MC modeling. The statistical uncertainties are taken into account for the Gaussian prior of the template fit (see Sec. 11.5). Note that the decay $Z \to \mu\mu$ and the decay of a W boson into two muons in single top processes do not give rise to $e \to \gamma$ fakes, as expected.

are tagged as prompt photons. Hadron fakes that are identified by this procedure are omitted since hadron fakes are already included by the hadron fake template in the final template fit.

Fig. 66 to Fig. 68 show the comparison of the distributions of photon track isolation between the prompt photon and hadron fake templates derived from data on the one hand and the photon isolation of the considered background events where the photons have been derived from the MC classification of the HERWIG decay chain information on the other hand; indicating that the photon models derived from data and from MC classification are in agreement within their statistical uncertainties.

Furthermore, truth electrons, mainly those originating from W and Z decays, might be reconstructed as photons. Hence, the photon is checked if it is located within a cone of $\Delta R < 0.15$ around the truth electron. If the photon is inside that cone, it is treated as an electron mis-identified as a photon, hence such contributions are multiplied with the corresponding $e \to \gamma_{\text{fake}}$ SF (see Sec. 11.4.1).

11 CROSS SECTION MEASUREMENT

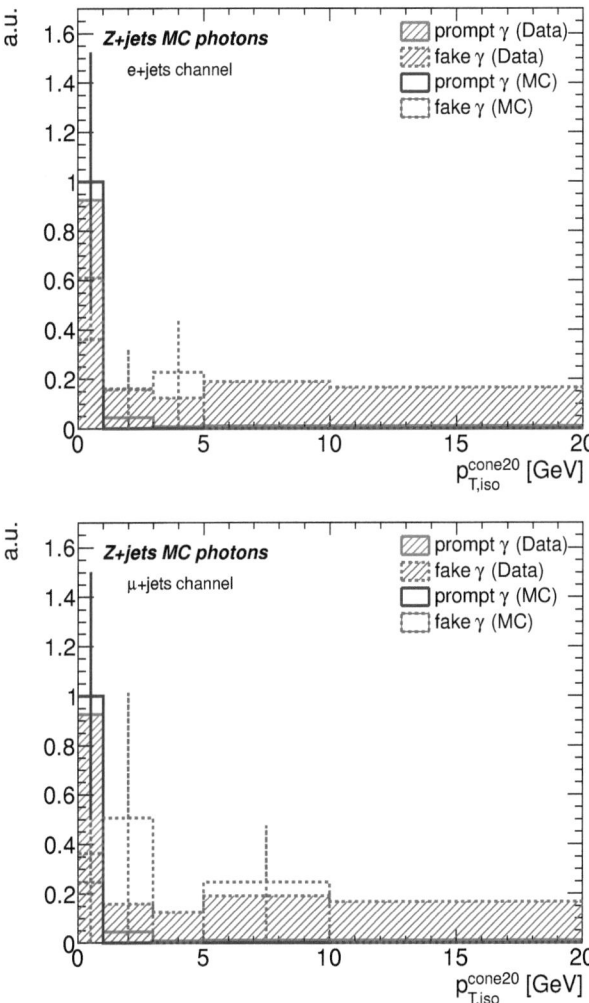

Figure 66: Z+jets photon isolation comparison of the prompt photon and hadron fake templates derived from data and the isolation spectra of photons derived from the classification of prompt photons and hadron fakes from the HERWIG decay chain information; depicted for the e+jets (upper plot) and μ+jets channel (lower plot). The error bars represent the statistical uncertainties.

11.4 Estimation of Prompt Photon Background Contributions

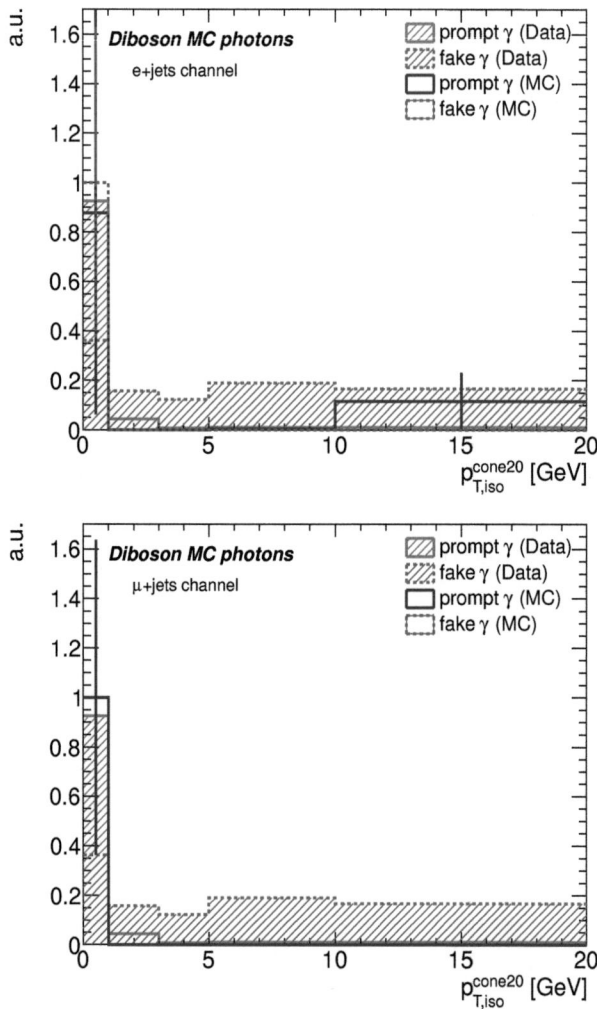

Figure 67: Di-boson photon isolation comparison of the prompt photon and hadron fake templates derived from data and the isolation spectra of photons derived from the classification of prompt photons and hadron fakes from the HERWIG decay chain information; depicted for the e+jets (upper plot) and μ+jets channel (lower plot). The error bars represent the statistical uncertainties.

11 CROSS SECTION MEASUREMENT 213

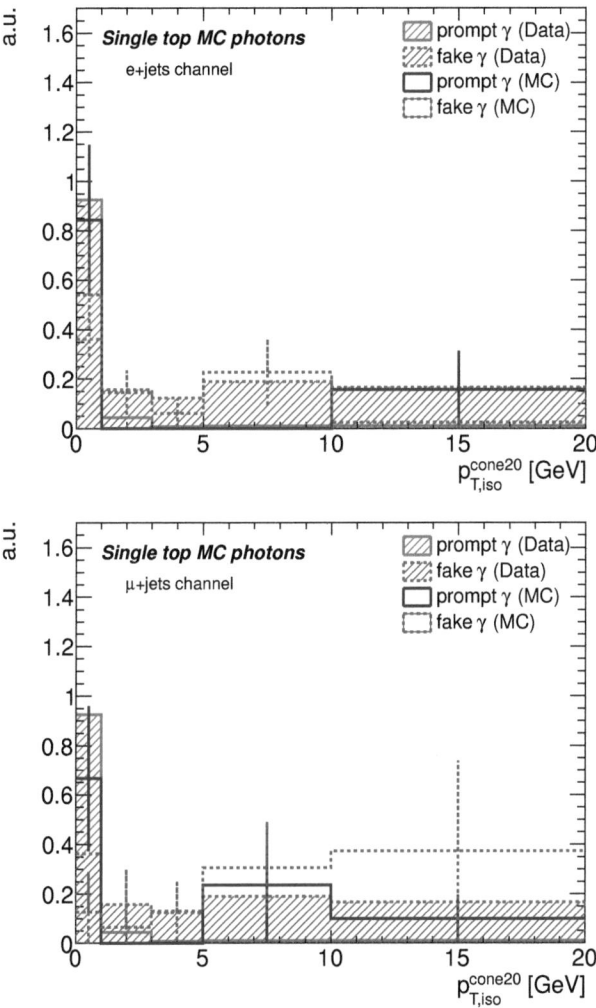

Figure 68: Single top quark photon isolation comparison of the prompt photon and hadron fake templates derived from data and the isolation spectra of photons derived from the classification of prompt photons and hadron fakes from the HERWIG decay chain information; depicted for the e+jets (upper plot) and μ+jets channel (lower plot). The error bars represent the statistical uncertainties.

11.5 Result of the Template Fit

The final template fit for obtaining the number of $t\bar{t}\gamma$ signal events is performed using the hadron fake template and prompt photon templates as described in Sec. 11.3.2 and Sec. 11.3.3. The maximum likelihood of the fit is calculated for the e+jets and the μ+jets channel together; the result of the expected $t\bar{t}\gamma$ signal already takes into account the estimated event selection efficiencies (see Sec. 10.8.1), yielding one common number of expected signal events for both channels. The event selection efficiency enters the template fit as a bare number devoid of any uncertainty[21].

Gaussian prior probabilities are assigned to the various prompt photon background sources according to their expectation (see Tab. 17), the width of the Gaussian prior is assigned the statistical uncertainty of the respective background contribution. The signal expectation and the hadron fake background are applied constant priors since there is no a priori estimation of these contributions; moreover, in case of the $t\bar{t}\gamma$ signal, the fit result should not be biased by any expectation or dedicated prior.

The prompt photon background contributions are grouped into classes of backgrounds; the yields of the di-boson, single top and Z+jets+γ contributions are combined in one number for the non-$t\bar{t}$ background, the amount of events where electrons have been mis-identified as photons ($e \to \gamma$ fakes) are composed from the according contributions of $t\bar{t}$, di-boson, single top and Z+jets+γ processes.

Fig. 69 shows the result of the final template fit for the e+jets (left) and the μ+jets channel (right). The template fit is performed simultaneously in both channels (cf. Eq. 142), the distribution of expected events is shown before the application of the event selection efficiencies.

---

[21]Systematic uncertainties that affect the event selection efficiency are investigated in Sec. 12.

11 CROSS SECTION MEASUREMENT

Contribution	Prior type	start value e+jets channel	start value μ+jets channel	fit result e+jets channel	fit result μ+jets channel
Signal ($t\bar{t}\gamma$)	constant	common fit parameter		1910 ± 510	
Hadron fakes	constant	—	—	20 ± 6	26 ± 7
$t\bar{t}$	Gaussian	0.25 ± 0.03	0.59 ± 0.07	0.25 ± 0.03	0.59 ± 0.07
W+jets+γ	Gaussian	3.6 ± 0.6	4.3 ± 0.5	3.5 ± 0.6	4.2 ± 0.5
$e \to \gamma$ fakes	Gaussian	8.2 ± 0.6	10.25 ± 0.04	8.2 ± 0.6	10.25 ± 0.04
Multijet BG	Gaussian	$0.39^{+0.40}_{-0.39}$	$0.24^{+0.25}_{-0.24}$	$0.38^{+0.40}_{-0.38}$	0.25 ± 0.25
other non-$t\bar{t}$ BG	Gaussian	2.16 ± 1.15	1.79 ± 1.13	2.09 ± 1.13	1.84 ± 1.12

Table 17: Relevant processes that are added to the template fit. The assigned prior type and the initial values taken from the various background contributions are shown either. Note that only the statistical uncertainties of the background expectations are quoted. The fit result of the $t\bar{t}\gamma$ signal already takes into account the event selection efficiency.

After marginalization, the expectation value of signal events $N_{\exp}(t\bar{t}\gamma)$ of the combined fit after the application of the event selection efficiencies reads

$$N_{\exp}(t\bar{t}\gamma) = 1880 \pm 510$$

where the statistical uncertainties are symmetrized ($\sigma = \sqrt{V}$). Looking at the 68 % central confidence interval around the expectation value, the result is $N_{\exp}^{68\%}(t\bar{t}\gamma) = 1870^{+520}_{-490}$. The marginalized distribution and the 68 % confidence interval are shown in Fig. 70. The result translates to a cross section of

$$\sigma_{t\bar{t}\gamma} \times \text{BR} = [1.84 \pm 0.49(\text{stat.})] \text{ pb}$$

for an integrated luminosity of $1.04\,\text{fb}^{-1}$. The overall background is estimated to $N_{\text{BG}} = 78.3 \pm 9.8(\text{stat.})$ in the e+jets and the μ+jets channel together, the prompt photon background (including electrons mis-identified as photons) yields $31.6 \pm 1.9(\text{stat.})$.

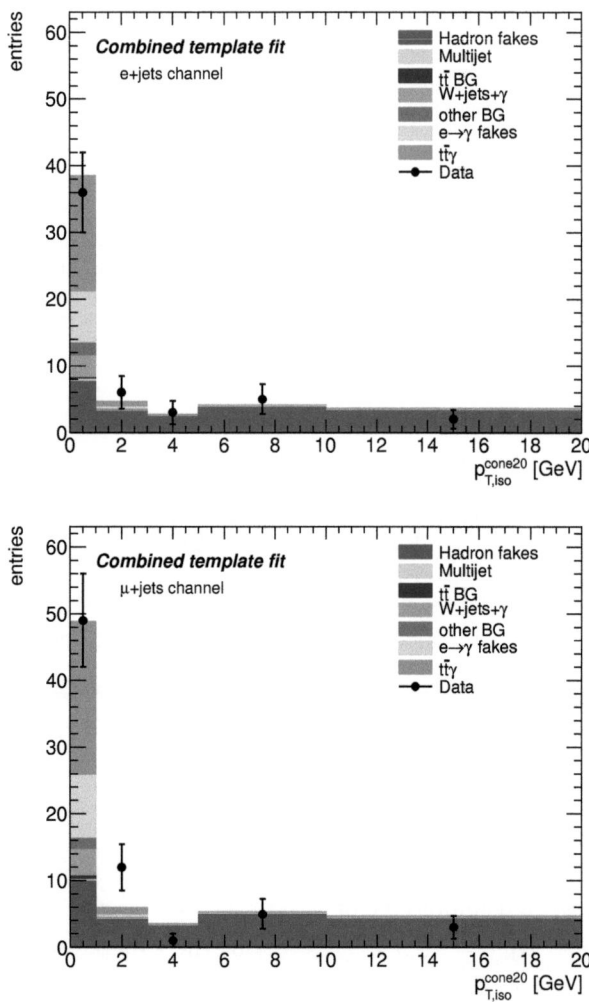

Figure 69: Result of the final template fit for the e+jets (upper plot) and the μ+jets channel (lower plot). The template fit is performed simultaneously in both channels. The number of entries in the histograms is given before the application of the event selection efficiencies.

11 CROSS SECTION MEASUREMENT

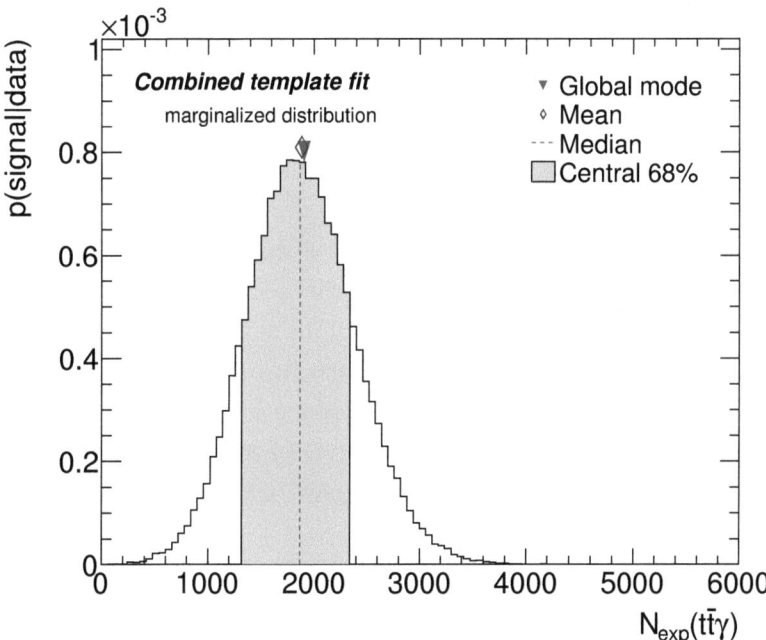

Figure 70: Marginalized distribution of expected number of $t\bar{t}\gamma$ candidates after the application of the event selection efficiencies (i.e. before event selection/detector acceptance) obtained from the combined template fit.

12 Systematic Uncertainties

The systematic uncertainties that affect the measurement of the $t\bar{t}\gamma$ cross section measurement are basically the same as for any standard $t\bar{t}$ cross section measurement, plus additional uncertainties that are related to photon reconstruction and identification. Besides, there are systematic uncertainties that affect the method of the measurement, such as the modeling of the prompt photon and hadron fake templates and the estimation of the various prompt photon background contributions.

Considering the standard $t\bar{t}$ uncertainties, there are two kinds of them: Some alter the event yield of the event selection due to shifts of the lepton or jet energy or the acceptance of objects (like the treatment of the LAr calorimeter hardware failure or jet reconstruction efficiency); others merely modify the event weight of the unaltered event selection, such as object scale factors (see Sec. 10.6.2).

This section describes all considered systematic uncertainties and the method of their determination as well as the method for obtaining the final systematic uncertainty of the $t\bar{t}\gamma$ cross section measurement.

12.1 Method

The impact of the systematic uncertainties is analyzed separately for each one, i.e. only one systematic uncertainty is considered at the same time, shifting its expectation up and down by one standard deviation $\sigma_{\text{syst.}}$, where $\sigma_{\text{syst.}}$ is defined as the expected effect of the uncertainty. In principle, the cross section measurement could then be just redone on data with the modified event selection efficiency, background estimations and templates, resulting in one altered cross section result.

Since the data events are subject to statistical fluctuations (Poissonian distribution), the effect of the systematic uncertainty might be strongly biased by the underlying statistical fluctuations.

12 SYSTEMATIC UNCERTAINTIES

Hence for each systematic uncertainty, 3000 template fits are performed using pseudo data derived from MC expectations (see Sec. 12.2), each one with each bin of the pseudo data histograms fluctuated according to the Poissonian uncertainty in that bin. This procedure yields a distribution of expected number of signal events N_{exp} with its maximum at the most likely isolation distribution in pseudo data. The mean of the N_{exp} distribution is assumed to be the most likely effect of the considered systematic uncertainty.

The same procedure (3000 pseudo experiments) is performed with no systematic uncertainty applied. The relative difference of the mean of that distribution of expected signal events $\bar{N}^0_{\text{exp}}$ and the pseudo experiment ensemble with systematic uncertainty considered ($\bar{N}_{\text{exp}}$) is then taken as systematic uncertainty of the cross section measurement:

$$\sigma_{\text{syst.}} = \frac{\bar{N}_{\text{exp}} - \bar{N}^0_{\text{exp}}}{\bar{N}^0_{\text{exp}}}. \qquad (153)$$

12.2 Pseudo Data

In order to avoid any statistical bias, pseudo data is created from MC expectations, since the number of data events surviving the $t\bar{t}\gamma$ event selection is a fixed outcome of the experiment and hence may not be altered anymore. Besides, most of the systematic uncertainties are only applicable for MC simulations and cannot be applied on data directly. Hence, a MC based expectation of the measured photon isolation shapes (*pseudo data*) is evaluated.

The pseudo data contains

- the $t\bar{t}\gamma$ signal expectation,
- all prompt photon background estimations,
- the expected amount of hadron fakes in each channel

and is created for the e+jets and the μ+jets channel separately.

The prompt photon background contributions have been estimated in Sec. 11.4.2 to Sec. 11.4.5. For each background source, the prompt photon template is scaled with the expected background yield and added to the pseudo data.

The expected amount of hadron fakes per channel has to be taken from the template fit performed in Sec. 11.5 because there is no other possible way for estimation. The template fit yields a hadron fake contribution of 20 ± 6 events in the e+jets channel and 26 ± 7 events in the μ+jets channel. The hadron fake template is normalized to these numbers and added to the pseudo data either.

Fig. 71 shows the generated pseudo data for the electron and the muon channel together with real data for comparison and indicates that both pseudo data and measured data are in good agreement.

Fig. 72 shows examplarily the result of 3000 pseudo experiments with the jet energy scale uncertainty varied, Fig. 73 illustrates the impact of the corresponding uncertainty due to b-tagging performance. The complete set of these plots for all considered systematic uncertainties is provided in Sec. C.

12.3 Sources of Systematic Uncertainties

This chapter describes all source of systematic uncertainties that have been taken into account.

12.3.1 Jet Modeling

The JES systematic uncertainty is determined using a combination of a data-driven analysis and the systematic variations of MC simulations [85].

The sources of contributions to the EM+JES systematic uncertainties are:

12 SYSTEMATIC UNCERTAINTIES

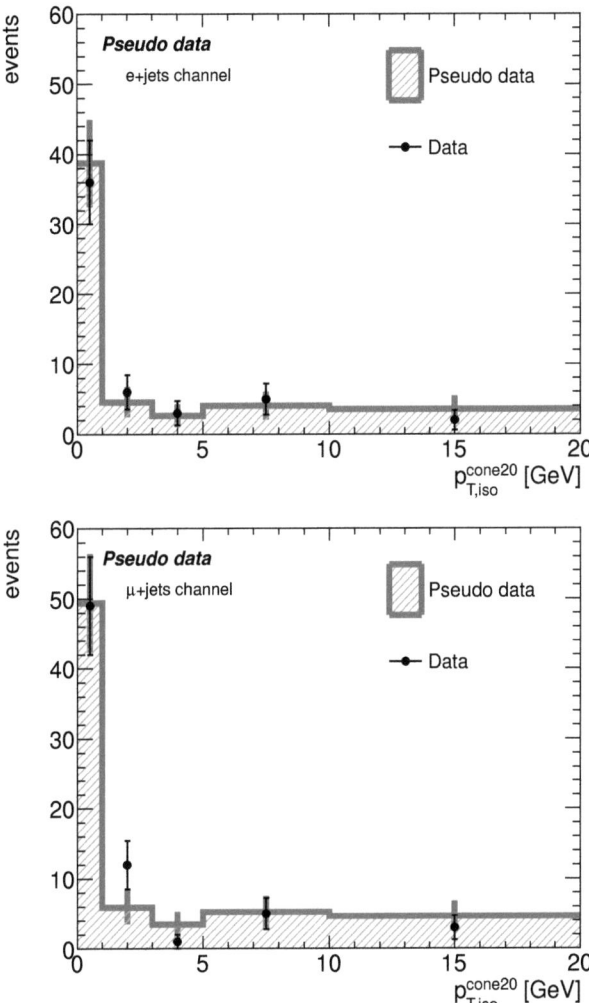

Figure 71: Pseudo data generated for the electron channel (upper plot) and the muon channel (lower plot). The error bars of pseudo data are the expected statistical uncertainties for the given number of expected events. The real data including its statistical uncertainty is shown for comparison. The expectation of pseudo data and real data are in good agreement.

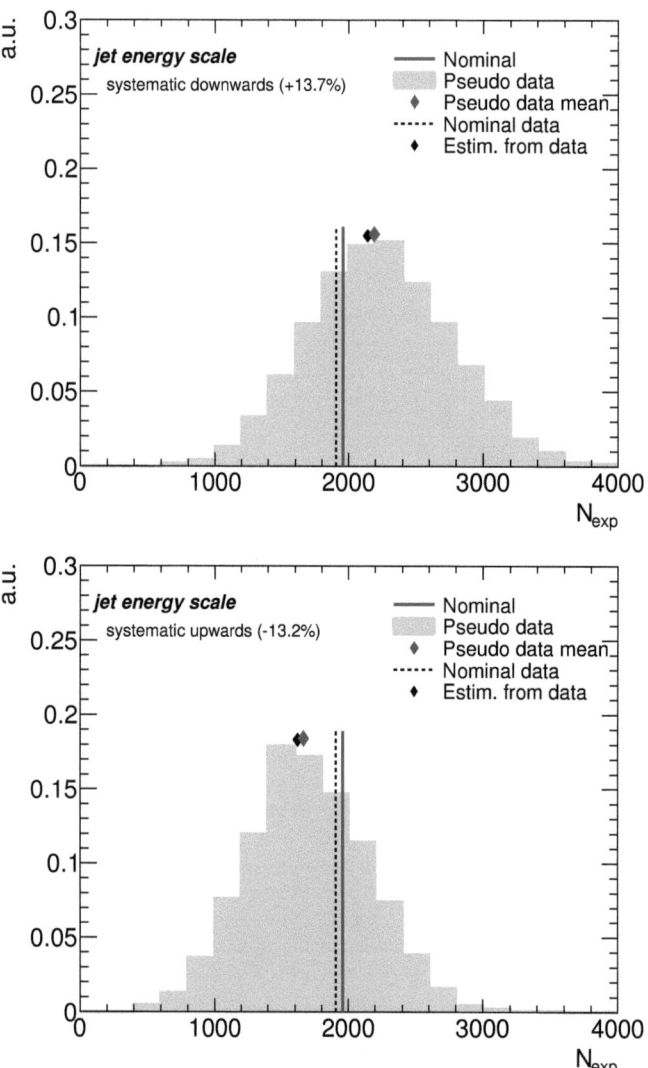

Figure 72: Distribution of the expected number of signal events N_{exp} for 3000 pseudo experiments, shown for the impact of the jet energy scale uncertainty (JES).

12 SYSTEMATIC UNCERTAINTIES 223

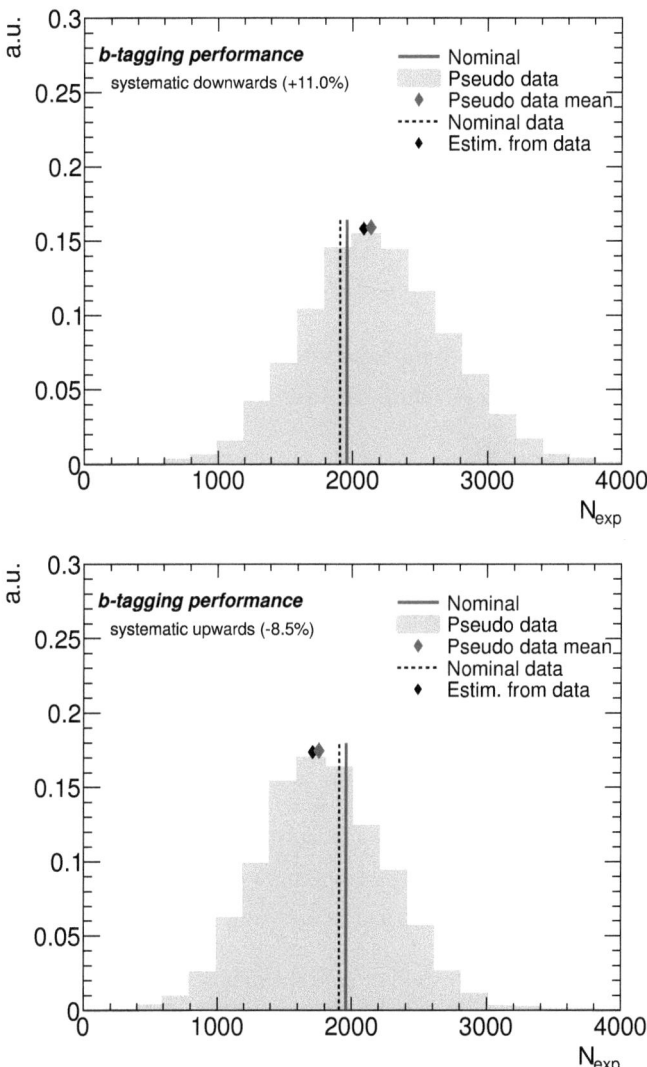

Figure 73: Distribution of the expected number of signal events N_{exp} for 3000 pseudo experiments, shown for the impact of the b-tagging performance.

Method of JES Calibration After the nominal inclusive jet Monte Carlo simulation sample is calibrated, the jet energy and p_T response still shows slight deviations from unity at low p_T (*non-closure*). Any deviation from unity (non-closure) in transverse momentum and energy response after the application of the JES to the nominal Monte Carlo sample implies that the kinematics of the calibrated calorimeter jet are not restored to that of the corresponding particle jets. This is mostly due to the following reasons:

- There is an underlying assumption that every constituent needs the same average compensation when deriving the calibration constants;

- The same correction factor for energy and transverse momentum is used. In the case of a non-zero jet mass that does not reflect the truth jet mass; restoring only the jet energy and pseudo-rapidity will lead to a bias in the p_T calibration.

Calorimeter Response The uncertainty of the calorimeter response to jets can be obtained from the response uncertainty of the individual particles constituting the jet using truth information:

- the single hadron energy measured in a cone around an isolated track with respect to the track momentum (E/p) in the momentum range from $0.5 < p < 20\,\text{GeV}$,

- the initial pion response measurements performed in the 2004 combined ATLAS test-beam, where a full slice of the ATLAS detector has been exposed to pion beams with momenta between 20 and $350\,\text{GeV}$.

Uncertainties for charged hadrons are estimated from these measurements. Additional uncertainties accounted for include:

- effects related to the calorimeter acceptance,

12 SYSTEMATIC UNCERTAINTIES 225

- uncertainties related to particles with large momenta of $p > 400\,\text{GeV}$,

- baseline absolute electromagnetic scale for the hadronic and electromagnetic calorimeters for particles not measured in-situ,

- uncertainties connected to neutral hadrons.

At high transverse momentum, the dominating contribution to the calorimeter response uncertainties is due to larger momenta particles.

Detector Simulation Topoclusters are constructed based on the signal-to-noise ratio of calorimeter cells. Discrepancies between the simulated noise and the real noise in data can lead to differences in the cluster shapes and to the presence of fake clusters, which affect the jet reconstruction. For data, the noise can change over time, while the noise RMS used in the simulation are fixed at the time of the production of the simulated data hence MC simulation does not reflect the noise in data.

Physics Modeling The contributions to the JES uncertainty from the modeling of the fragmentation and underlying event and other parameters of the Monte Carlo event generator are obtained using different Monte Carlo samples, fragmentation algorithms and tuning of parameters that control parton showering.

Pseudo-Rapidity Extrapolation The JES uncertainty determined in the central detector is extrapolated to the forward regions by exploiting the transverse momentum balance of a central and a forward jet in events with di-jet topologies. In such events, the responses of the forward jets are measured relative to those of the central jets.

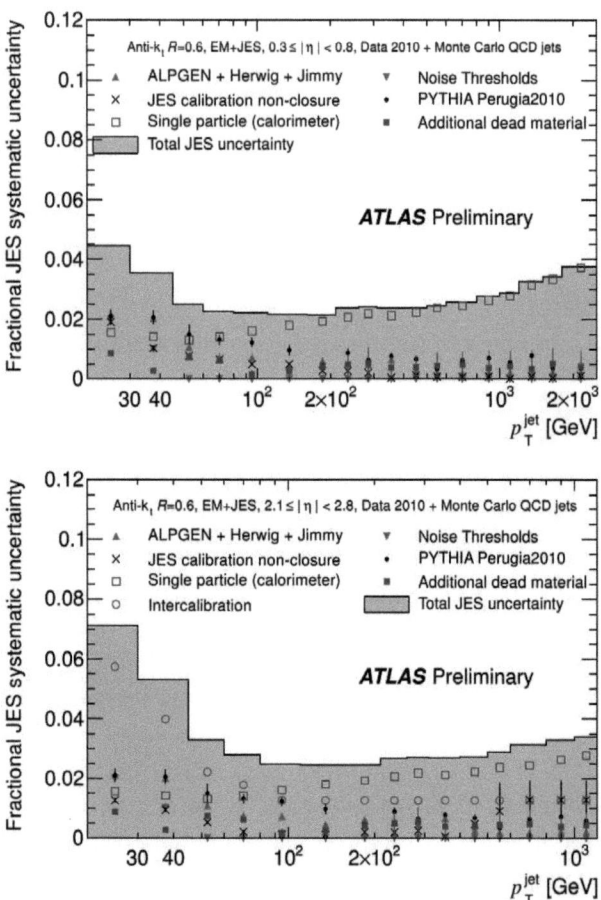

Figure 74: Summary plots of JES uncertainties for the central (upper plot) and the end-cap detector region (lower plot) [85].

Fig. 74 shows the summary of the considered JES uncertainties estimated for the central and the end-cap detector region.

The resolution of the measured jet energy (JER) was measured from 2010 data [144].

12.3.2 b-Tagging Performance

The p_T^{rel} spectrum of muons from b-jet decays (see Sec. 7.3.5) is estimated from template fits considering different flavor compositions. These templates have an intrinsic statistical uncertainty and have been varied taking into account the following effects [91]:

- the difference of the direction of the B hadron in $\eta - \phi$ space between data and MC simulation,

- contamination of the light flavor control sample with b-jets,

- the modeling of b and c quark production,

- the modeling of B hadron decays and b quark fragmentation with PYTHIA,

- the difference of muon p_T spectra between data and MC simulation,

- artificial suppression of fake muons w.r.t. muons from B decays due to a p_T cut on muons at generator level,

- b-tag efficiency scale factors,

- pileup effects.

The overall b-tagging uncertainties are at the order of 13 % and slightly depend on the jet p_T.

12.3.3 $\not{E}_T$ Uncertainties

As described in Sec. 7.5, the $\not{E}_T$ is calculated from all reconstructed physical objects and additional energy deposited in cluster cells that do not belong to objects (cell-out term).

Since the measured energy of all reconstructed objects are subject to systematic uncertainties, their energy variation will have an impact on the value of $\not{E}_T$ as well as the uncertainty of measured energy of the cell-out contribution.

The systematic uncertainties in the cell-out term have been estimated using dedicated MC samples and from topocluster energy scale uncertainties [97]. The main sources are discrepancies in the imperfect knowledge of the distribution of material in the detector, the choice of the parton shower model and the modeling of the underlying event. The topocluster method yields the largest uncertainty of $\approx 13\%$.

The same two methods have been used to estimate the systematic uncertainties of the soft-jet term in $\not{E}_T$, yielding an uncertainty of 10.5 %.

The effect of pile-up is taken into account by an uncertainty of 10 % and is applied on the cell-out and soft jet term.

12.3.4 Electron, Photon and Muon Performance

The discrepancy of trigger, ID and reconstruction efficiencies between data and MC simulation for electrons and muons is corrected by scale factors (see Sec. 10.6.1). The SFs have systematic uncertainties at the order of 3 % (0.5 %, 1 %) for the identification (trigger, reconstruction) performance respectively, depending on the pseudo-rapidity and transverse energy of the electron [136].

The muon reconstruction (trigger) SFs have uncertainties of 0.3 % (0.6...18 %) respectively. The uncertainty on the muon identification SF is assumed to be negligibly small.

Energy Resolution The finite energy resolution in the EMC is translated to electron and photons in MC simulations by smearing their nominal energy by a random value taken from a Gaussian distribution with its width assigned the energy resolution of the detector. The energy resolution of combined muons depends on both the energy resolution in the MS as well as in the ID. Hence, the effect of smearing the measured energy of muons in the MS and in the ID is translated to the transverse momenta reconstructed in the MS, the ID and combined muons. The maximum relative uncertainty of all combinations is taken as systematic uncertainty.

Energy Rescaling The scaling of the energies of electrons and photons is rescaled as described in Sec. 9.5. The systematic uncertainties of the energy rescaling are in the range from $-1.5\,\%$ to $2\,\%$.

12.3.5 LAr Hardware Failure

The decision whether to reject an MC event or not due to the emulation of the LAr hardware failure considers jets situated near the regions of the EMC where six of FEBs failed after a technical stop and depends on a p_T threshold (see Eq. (135) in Sec. 9.8). This threshold is varied by $\pm 4\,\text{GeV}$, giving rise to a different rejection of jets and hence different event yields.

12.3.6 PDF Uncertainty

The generation of simulated events strongly depends on the momenta of the incoming particles. The distribution of these momenta is determined by the choice of the PDF (see Sec. 2.2.1).
The distribution of the momenta of the two colliding particles depends on the intrinsic uncertainties of the measurement of one specific PDF (*intra-PDF uncertainty*) and on the choice of the PDF (*intra-PDF uncertainty*), since PDF sets are provided by different research groups, using the experimental results of different experiments.
For the evaluation of the systematic uncertainties of the $t\bar{t}\gamma$ cross section measurement that emerge from PDF uncertainties, three different sets of PDF are considered:

1. **CT10**: This PDF set is measured and extrapolated at the NLO level [145]. It contains one nominal PDF set and 52 eigenvector

sets. The strong coupling constant $\alpha_s(M_Z)$ is varied in the range from 0.116 to 0.120 and from 0.112 to 0.127.

2. **MSTW2008nlo68cl**: This PDF set is extrapolated at the NLO level and contains 42 error PDF sets at the 68% confidence level [33] [146].

3. **NNPDF20_100**: This error set contains 100 error PDF sets [147].

For each event, a new probability weight (*PDF weight* w_{PDF}) is calculated from the original momentum fractions of the incoming partons $f_0(x_1, Q^2)$ and $f_0(x_2, Q^2)$ and the according fractions calculated from the new PDF set $f_1(x_1, Q^2)$ and $f_1(x_2, Q^2)$ by

$$w_{\text{PDF}} = \frac{f_1(x_1, Q^2) \cdot f_1(x_2, Q^2)}{f_0(x_1, Q^2) \cdot f_0(x_2, Q^2)}. \tag{154}$$

Each of the PDF sets described above provides a set of error PDFs that can be used to derive the uncertainty of the corresponding PDF. A set of w_{PDF} is hence obtained for each of these error sets by Eq. (154). From the weight distributions of each PDF, their uncertainties are calculated. For CT10, the symmetric Hessian form is chosen, which is defined as

$$\sigma_{\text{CT10}} = \frac{1}{2}\sqrt{\sum_i (w^+_{\text{PDF},i} + w^-_{\text{PDF},i})^2}. \tag{155}$$

For MSTW2008nlo68cl, the asymmetric Hessian form is calculated which is defined as

$$\begin{aligned}
\sigma^+_{\text{MSTW}} &= \sqrt{\sum_i (w_{\text{PDF},i} - w_{\text{PDF},0})^2} & \text{if } w_{\text{PDF},i} - w_{\text{PDF},0} > 0, \\
\sigma^-_{\text{MSTW}} &= \sqrt{\sum_i (w_{\text{PDF},i} - w_{\text{PDF},0})^2} & \text{if } w_{\text{PDF},i} - w_{\text{PDF},0} < 0.
\end{aligned} \tag{156}$$

12 SYSTEMATIC UNCERTAINTIES

NNPDF does not provide a dedicate set of error PDFs. It provides an ensemble of equitable PDFs with parameters varied. Hence, the uncertainty of the NNPDF set is defined as the standard deviation:

$$\sigma_{\text{NNPDF}} = \sqrt{\sum_i (w_{\text{PDF},i} - \bar{w}_{\text{PDF}})^2} \qquad (157)$$

where $\bar{w}_{\text{PDF}}$ is the average of all NNPDF sets:

$$\bar{w}_{\text{PDF}} = \frac{1}{N-1} \sum_i^N w_{\text{PDF},i}. \qquad (158)$$

The PDF weights are obtained event by event and multiplied on each event weight. Thus, after a certain step in the event selection, an ensemble of event selection efficiencies can be extracted for the three considered PDFs and for all of their error PDFs. From the varied selection efficiencies, the envelope σ_{all} of all PDF uncertainties is then taken to be the overall uncertainty related to the PDF modeling. The envelope is defined as half of the difference between the minimum and maximum value of all uncertainties:

$$\sigma_{\text{all}} = \frac{\max(\Delta^+_{\text{CT10}}, \Delta^+_{\text{MSTW2008}}, \Delta^+_{\text{NNPDF}}) - \min(\Delta^-_{\text{CT10}}, \Delta^-_{\text{MSTW2008}}, \Delta^-_{\text{NNPDF}})}{2} \qquad (159)$$

with $\Delta w^+ = w_{\text{PDF},0} + \sigma$ and $\Delta w^- = w_{\text{PDF},0} - \sigma$ for the considered PDF set (or, in case of MSTW2008nlo68cl, $\Delta w^+ = w_{\text{PDF},0} + \sigma^+$ and $\Delta w^- = w_{\text{PDF},0} - \sigma^-$ respectively).

Since the three probe PDFs are all evaluated at the NLO level, the PDF weighting technique described here can only be applied to MC samples where a NLO PDF set was used during event generation. Hence, the PDF uncertainty can only be derived for the NLO MC@NLO $t\bar{t}$ sample applying the event preselection (i.e. without photon requirement) and is then assumed to be at the same order as for the LO

WHIZARD $t\bar{t}\gamma$ sample and the full event selection. The systematic uncertainties related to the PDF uncertainties read 1.3 % in the e+jets and 1.1 % in the μ+jets channel.

The distribution of the event selection efficiencies of the MC@NLO $t\bar{t}$ sample, varied by the PDF weights w_{PDF} for all error PDF sets, their uncertainties and the envelope of all uncertainties is illustrated in Fig. 75 for the electron and muon channel separately.

12.3.7 $t\bar{t}$ MC Modeling

The uncertainties due to the modeling of $t\bar{t}\gamma$ MC simulation are estimated from various $t\bar{t}$ samples produced with different MC generators and PS programs (cf. Tab. 31-33). The uncertainties are evaluated for the event preselection (see Sec. 9.9) and extrapolated to both the $t\bar{t}$ and $t\bar{t}\gamma$ event yields after the full event selection. The effects taken into account are described below, Tab. 18 shows a summary of all $t\bar{t}$ uncertainties. The sum of all uncertainties is applied to the background estimations where contributions of $t\bar{t}$ or $t\bar{t}\gamma$ are involved and thus affect the estimated uncertainty of the background contributions. Additionally, all single uncertainties are applied separately for the estimation of the final systematic uncertainty of the $t\bar{t}\gamma$ cross section measurement (see Sec. 12.4).

For the evaluation of the $t\bar{t}$ uncertainties, only the bare MC weights have been taken into account instead of the complete event weights (cf. Sec. 10.6.2) in order to disentangle the evaluation of the uncertainties of the $t\bar{t}$ MC modeling from uncertainties of scale factors.

Choice of the NLO MC Generator NLO calculations of the $t\bar{t}$ cross section and event generation can be performed using either the MC@NLO or the POWHEG generator. The differences of event yields resulting from the choice of MC@NLO as standard generator and the POWHEG sample using the same PS program as MC@NLO (HER-

12 SYSTEMATIC UNCERTAINTIES

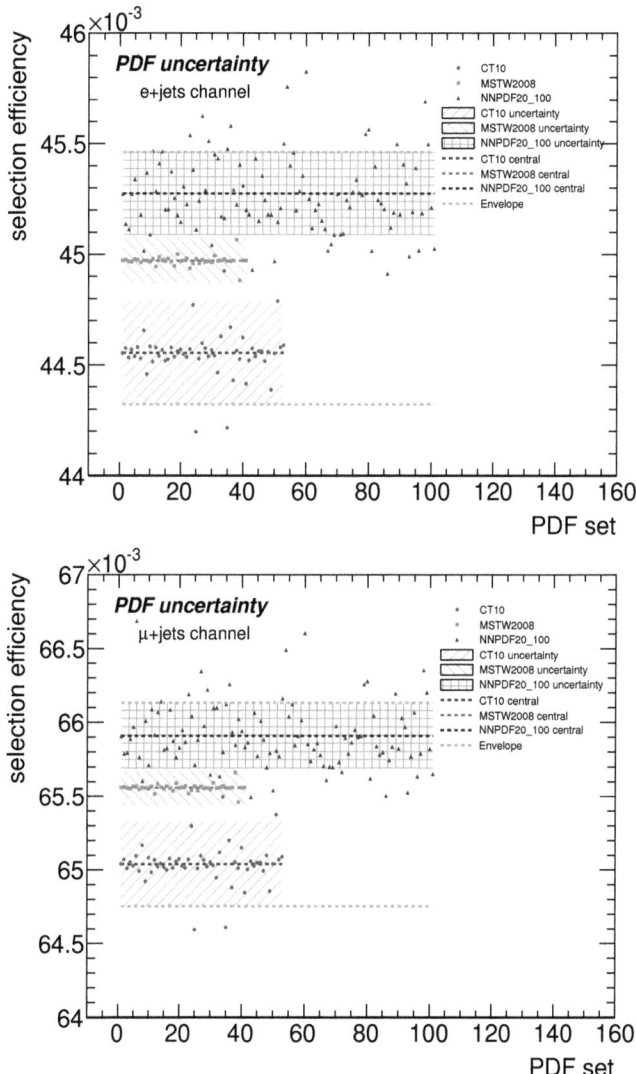

Figure 75: Illustration of the determination of the overall PDF uncertainty, evaluated from the MC@NLO $t\bar{t}$ sample. The distribution of varied event selection efficiencies (after the $t\bar{t}\gamma$ event preselection) from the PDF weights is shown for the electron (upper plot) and the muon channel (lower plot) separately.

	e+jets channel	μ+jets channel
ISR/FSR variations	6.9 %	5.8 %
Parton showering	1.7 %	1.8 %
LO vs. NLO	6.9 %	4.4 %
NLO generator	6.2 %	5.2 %
Sum	11.7 %	9.1 %

Table 18: Breakdown of the uncertainties related to $t\bar{t}$ MC modeling. The sum of all uncertainties is applied to the background estimations where contributions of $t\bar{t}$ or $t\bar{t}\gamma$ are involved.

WIG) is quoted as systematic uncertainty and yields 6.2 % (5.2 %) in the e+jets (μ+jets) channel.

NLO vs. LO Simulation The impact of the choice of a fixed order calculation at the LO and the NLO level is obtained by comparing the event yields of the MC@NLO $t\bar{t}$ sample with the yield of an AcerMC $t\bar{t}$ sample (LO) after the event pre-selection. Both samples have been showered with HERWIG.
The event yields give rise to a relative difference of 6.9 % (4.4 %) in the e+jets (μ+jets) channel.

Choice of the Parton Showering Program The impact of the choice of a different parton showering model has been investigated by comparing two POWHEG $t\bar{t}$ MC samples; one showered with HER-WIG, the second one showered with PYTHIA. Comparing the event yields of both samples leads to a relative difference of 1.7 % (1.8 %) in the e+jets (μ+jets) channel.

ISR and FSR Variations ISR and FSR radiation can lead to the production of additional or less jets and hence will affect the event yield. For the generation of AcerMC LO $t\bar{t}$ MC events, six additional samples with different settings of ISR/FSR radiation in the PYTHIA PS have been produced [148]:

12 SYSTEMATIC UNCERTAINTIES

- enhanced production of ISR radiation (PARP(67)=6.0 and PARP(64)=0.25) and reduced production of ISR radiation (PARP(67)=0.5 and PARP(64)=4.0),

- enhanced production of FSR radiation (MSTP(3)=1, PARP(72)=0.384 and PARJ(82)=0.5) and reduced production of FSR radiation (MSTP(3)=1, PARP(72)=0.096 and PARJ(82)=2.0),

- simultaneous enhancement and reduction of ISR and FSR radiation. The according parameters for steering ISR and FSR radiation described above are varied at the same time and set to the corresponding values.

The relevant parameters being modified for the ISR and FSR variations have the following meaning [75]:

- PARP(67): factor that is multiplied on the Q^2 scale of hard scattering and defines the maximum parton virtuality in Q^2-ordered space-like parton showers.

- PARP(64): factor that is multiplied on $k_\perp^2$ in space-like parton shower evolution.

- PARP(72): Λ value (in GeV) used in running α_s for time-like parton showers.

- MSTP(3): setting of this value to 1 adopts PARP(72) as Λ value in case of time-like parton showers.

- PARJ(82): Invariant mass cut-off (in GeV) below which PYTHIA assumes parton showers not to radiate anymore.

The maximum deviation of the event yields of six ISR/FSR variations is quoted as systematic uncertainty and yields 6.9% (5.8%) in the e+jets (μ+jets) channel.

12.3.8 $t\bar{t}\gamma$ NLO Calculations

The $t\bar{t}\gamma$ MC sample has been generated at the LO level using the WHIZARD MC generator. k-factors for the NLO calculation have been calculated for two different scenarios for a CMS energy of $\sqrt{s} = 7\,\text{TeV}$: Once for a renormalization energy scale of $\mu_R = m_t$ and once for $\mu_R = 2m_t$ [122].
The k-factor for $\mu_R = m_t$ reads $k(\mu_R = m_t) = 2.11$ and $k(\mu_R = 2m_t) = 2.55$ respectively. The k-factors were calculated for a running factorization scale taking into account the partonic CMS energy $\mu_F = \sqrt{\hat{s}}$.
The uncertainty of the $t\bar{t}\gamma$ cross section is assumed to be half the envelope of both k-factors and yields $k_{t\bar{t}\gamma} = 2.33 \pm 0.22$ which corresponds to a relative uncertainty of $9.4\,\%$.
The uncertainty of the $t\bar{t}\gamma$ cross section has been included in the estimations of the various prompt photon background sources, where needed.

12.3.9 Uncertainties from Prompt Photon Background Estimations

The systematic uncertainties of the various prompt photon background contributions have been evaluated together with the background estimations in Sec. 11.4.2 to Sec. 11.4.5.
The systematic uncertainty on the estimation of the W+jets+γ background is dominated by the uncertainties of the Berends-Giele scaling and the fit uncertainties of the estimation of the prompt photon fraction in the control region.
The uncertainty on the amount of multijet plus prompt photon background (QCD+γ) is determined by the global $100\,\%$ uncertainty on the normalization (see Sec. 10.7).
The $t\bar{t}$ background is applied the uncertainty due to the MC@NLO $t\bar{t}$ expectation (see Sec. 12.3.7).

The remaining non-$t\bar{t}$ background source are applied the uncertainties due to Berends-Giele scaling (Z+jets+γ) and the uncertainties of the single top and di-boson normalization (see Sec. 10.7).

The uncertainties of the estimation of the number of electrons misidentified as photons are included in the corresponding estimations of the $e \to \gamma$ fake contributions from $t\bar{t}$ processes and according contributions from di-boson, single top and Z+jets+γ processes.

12.3.10 Template Modeling

The prompt photon template derived from electron isolation using $Z \to ee$ decays (see Sec. 11.3.2) is applied a small correction (Eq. (142)).

Since this correction is estimated from MC simulation, an uncertainty of 100 % is applied by once not applying the correction to the isolation template and once applying it twice.

For the variation of the hadron fake template, the uncertainties of the exponential fit to the p_T spectrum (see Fig. 56 in Sec. 11.3.3) and the statistical uncertainties of the estimation of the $|\eta|$ spectrum have been taken into account. Furthermore, the p_T fit and the $|\eta|$ spectrum have been derived from MC simulation, once using the hadron fake photon definition (Sec. 11.1) and once using the HERWIG decay chain information. The exponential p_T fit has two free fit parameters: one for the normalization and one for the slope of the exponential curve. For the p_T fit, two variations of these parameters are considered: both being shifted upwards by there fit uncertainties and once both shifted downwards. This simultaneous shifting up and down leads to the largest impact on the integral of the curve.

For the variation of the hadron fake template hence exist $3 \times 3 \times 3 = 27$ combinations: Three possibilities for the $|\eta|$ weighting (data plus two MC models), the same three possibilities for the p_T spectrum and three possibilities to vary the p_T fit (nominal plus both fit parameters shifted upwards/downwards).

Each of these 27 combinations has been performed a χ^2-test on the nominal hadron fake template; the combination that yields the smallest p-value is taken as the systematic variation of the hadron fake template. Note that this procedure makes the distinction between a systematic shift upwards and a shift downwards meaningless. Hence, there is only one systematic variation of the hadron fake template.

Fig. 76 shows the comparison between the nominal and the systematic variation(s) of the prompt photon and the hadron fake template.

12.3.11 Possible Uncertainties due to Pile-Up Effects

Charged tracks or additional entries in calorimeter cells from pile-up events might distort the reconstruction the jet energy, $\not{E}_T$ or have an effect on the isolation of reconstructed leptons.

The event selection efficiencies for all average number of bunch crossings $\langle n_{BX} \rangle$ considered in the $t\bar{t}\gamma$ MC signal sample have been evaluated in order to study a possible impact of pile-up effects.

Fig. 77 shows the distribution of event selection efficiencies in dependence of $\langle n_{BX} \rangle$ as simulated in the $t\bar{t}\gamma$ MC signal sample. A constant fit is performed within the relevant range of $\langle n_{BX} \rangle$ values that are present in the analyzed data. The fits are within the statistical uncertainties of the considered efficiencies, indicating that there is no systematic behavior due to pile-up effects.

12.4 Combination of Uncertainties

The largest effect of all considered systematic uncertainties becomes visible in the variation of the event selection efficiency which is estimated from $t\bar{t}\gamma$ MC simulation. The efficiency has been recalculated

12 SYSTEMATIC UNCERTAINTIES

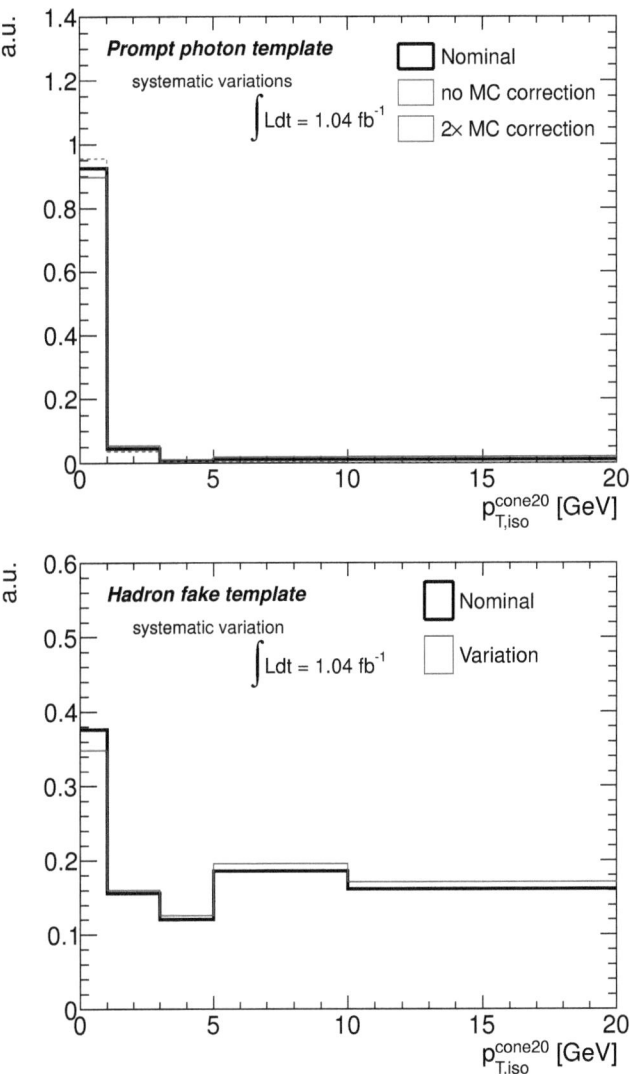

Figure 76: Systematic variations of the prompt photon template (upper plot) and the hadron fake template (lower plot).

for each of the systematic uncertainties described above, if applicable, and the final effect on the signal expectation is quoted as uncertainty. Although the estimated background contributions are estimated from different event selections, their actual yield is the outcome of the $t\bar{t}\gamma$ signal event selection. Hence, the relative difference of the selection efficiency w.r.t. the considered uncertainty is added to the expected background yield as well. The estimation of the amount of events from electrons mis-identified as photons is an exception, since the uncertainties of the trigger, reconstruction and ID SFs of electrons are already taken into account in the $e \rightarrow \gamma$ fake rate estimation as described in Sec. 11.4.1; hence this background contribution remains unchanged when considering the uncertainties of the electron SFs.

The uncertainties of the several background contributions are studied with the event selection efficiency remaining unchanged, the same holds for the study of the effect of the variation of the prompt photon and hadron fake templates.

The final uncertainties of the number of estimated signal events $N_{\text{exp.}}(t\bar{t}\gamma)$ are quoted as asymmetric errors. Since the results of the 3000 pseudo experiments are widely spread around the average value, the estimated uncertainties have a relatively large intrinsic statistical uncertainty at the order of 0.5 %. Hence, systematic uncertainties below that value cannot be resolved and are quoted as the statistical uncertainty of the pseudo experiment ensemble, being the upper limit. Tab. 19 shows the breakdown of all considered systematic uncertainties. The relative overall systematic uncertainty for the estimation of the background yield is smaller and reads $\sigma_{\text{BG}}^{\text{syst.}} = {}^{+8.5}_{-8.9}$ %.

12 SYSTEMATIC UNCERTAINTIES 241

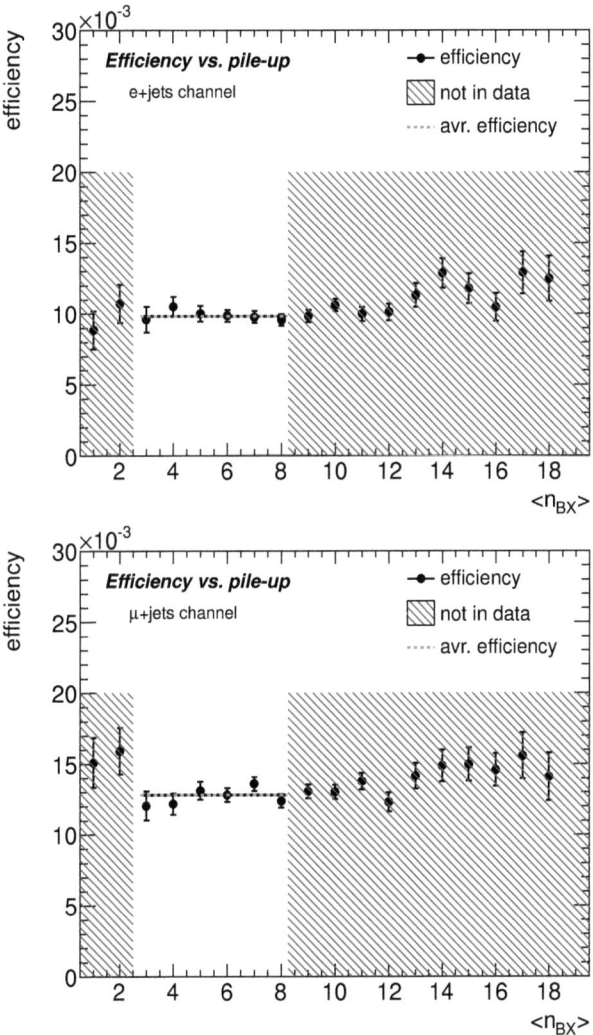

Figure 77: Event selection efficiency of the $t\bar{t}\gamma$ signal sample evaluated for different numbers of average bunch crossings $\langle n_{\mathrm{BX}} \rangle$, shown for the e+jet channel (upper plot) and the μ+jets channel (lower plot). A constant fit is performed within the relevant range of $\langle n_{\mathrm{BX}} \rangle$ present in the analyzed data. The fits are within the statistical uncertainties of the considered efficiencies, indicating that there is no systematic behavior due to pile-up effects.

Source of uncertainty	uncert. downwards ($\sigma^-_{\text{syst.}}$)	uncert. upwards ($\sigma^+_{\text{syst.}}$)
Jet performance	**−17.0 %**	**+16.8 %**
Jet energy scale (JES)	−13.7 %	+13.2 %
Jet energy scale incl. pile-up effects	−5.2 %	+5.6 %
b-jet energy scale (bJES)	−1.7 %	+2.2 %
Jet energy resolution (JER)	−2.2 %	+2.4 %
Jet reconstruction efficiency	−8.1 %	+8.1 %
b-tagging performance	**−8.5 %**	**+11.0 %**
Muon performance	**−3.0 %**	**+2.8 %**
Muon reconstruction efficiency	≤ 0.5 %	≤ 0.5 %
Muon ID efficiency	≤ 0.5 %	≤ 0.5 %
Muon trigger efficiency	−2.2 %	+2.2 %
Muon energy resolution	−1.9 %	+1.6 %
Muon energy resolution in the MS	−1.9 %	+1.6 %
Muon energy resolution in the ID	−1.6 %	+1.9 %
Muon energy scale	−0.9 %	≤ 0.5 %
Electron performance	**−1.9 %**	**+1.9 %**
Electron reconstruction efficiency	≤ 0.5 %	≤ 0.5 %
Electron ID efficiency	−1.6 %	+1.6 %
Electron trigger efficiency	≤ 0.5 %	≤ 0.5 %
Electron energy resolution	≤ 0.5 %	≤ 0.5 %
Electron energy scale	≤ 0.5 %	≤ 0.5 %
Photon performance	**−8.4 %**	**+7.7 %**
Photon ID efficiency	−8.4 %	+7.7 %
Photon energy resolution	≤ 0.5 %	≤ 0.5 %
Photon energy scale	≤ 0.5 %	+0.9 %
Missing transverse energy ($\not{E}_T$)	**−0.7 %**	**+0.7 %**
Pile-up effects on $\not{E}_T$	≤ 0.5 %	≤ 0.5 %
$\not{E}_T$ cell-out term	≤ 0.5 %	≤ 0.5 %
LAr hardware failure	**−2.5 %**	**+1.5 %**
MC uncertainties	**−10.7 %**	**+9.7 %**
PDF uncertainties	−1.1 %	+1.1 %
Choice of NLO generator	−5.9 %	+5.3 %
LO vs. NLO calculation	−5.7 %	+5.2 %
Modeling of parton shower/hadronization	−1.7 %	+1.7 %
ISR/FSR variations	−6.6 %	+5.9 %
Background modeling	**−8.2 %**	**+8.3 %**
W+jets+γ background	−5.0 %	+5.0 %
Electron mis-identification ($e \to \gamma$ fakes)	−5.5 %	+5.6 %
$t\bar{t}$ background	≤ 0.5 %	≤ 0.5 %
Multijet background	−1.3 %	+1.4 %
remaining non-$t\bar{t}$ background	−3.1 %	+3.1 %
Template modeling	**−9.8 %**	**+10.4 %**
Prompt photon template	−8.4 %	+9.1 %
Hadron fake template	−5.0 %	+5.0 %
Sum	**−27.0 %**	**+27.3 %**
Symmetrized	±27.2 %	

Table 19: Breakdown of the final systematic uncertainties of the $t\bar{t}\gamma$ cross section measurement.

13 Results

The template fit performed in Sec. 11.5 yields a cross section of $\sigma_{t\bar{t}\gamma} \times \mathrm{BR} = (1.89 \pm 0.48)\,\mathrm{pb}^{-1}$. Together with the systematic uncertainties of $^{+27.3}_{-27.0}\,\%$ evaluated in Sec. 12 and a global uncertainty of $\pm 3.7\,\%$ on the integrated luminosity [149], the final result for the $t\bar{t}\gamma$ cross section, branching into the semi-leptonic and di-leptonic decay mode, reads:

$$\sigma_{t\bar{t}\gamma} \times \mathrm{BR} = \left[1.84 \pm 0.49(\mathrm{stat.})^{+0.50}_{-0.50}(\mathrm{syst.}) \pm 0.07(\mathrm{lumi.})\right]\,\mathrm{pb}\,.$$

This result is in good agreement with the theoretical prediction of $\sigma^{\mathrm{MC}}_{t\bar{t}\gamma} = (1.96 \pm 0.18)\,\mathrm{pb}$, assuming a k-factor of 2.33 ± 0.22.

13.1 Significance Check

The significance of the result has been tested against the background-only hypothesis, i. e. the probability of measuring N_{data} (or even more) events in data for a given background estimation N_{BG} is evaluated.
In the e+jets and the μ+jets channel together, 122 events have been selected. The background estimation of the template fit (see Sec. 11.5) yields $78.3 \pm 9.8(\mathrm{stat.})^{+6.7}_{-7.0}(\mathrm{syst.})$ in both channels together.
The background expectation with the given fit and systematic uncertainties has been modeled by the convolution of a bifurcated Gaussian with a Poissonian distribution in order to account for additional statistical fluctuations of the actual number of background events in data. A bifurcated Gaussian has been used since the systematic uncertainties of the background estimation are asymmetric in general.
The probability to measure at least 122 background events at the complete absence of a signal contribution is $p = 0.176\,\%$. Considering a standard normal Gaussian distribution, this p-value translates to a significance of $2.9\,\sigma$.

The significance check has been performed for each of the 3000 pseudo experiments (see Sec. 12.1) for the background and data expectation of pseudo data.
The average result of the ensemble of pseudo experiments is $(3.1 \pm 0.9)\,\sigma$, so both the measured and the expected significance are in good agreement.
Fig. 78 shows the result for the significance of the measurement and the expected significance estimated from 3000 pseudo experiments.

13.2 Discovery Potential

The significance of having measured at least some $t\bar{t}\gamma$ signal by rejecting the background-only hypothesis depends both on the amount of analyzed data and the size of systematic uncertainties. Assuming that a physical process has not been discovered unless a level of $5\,\sigma$ will be reached, it is interesting to estimate when this will happen presumably.

The expected number of events is linearly scaled by the ratio of integrated luminosities L/L_0 ($L_0 = 1.04\,\text{fb}^{-1}$), the fraction of expected background is retained; the fit uncertainty of the background estimation is multiplied with an additional factor of $\sqrt{L/L_0}$ in order to account for an increasing Poissonian uncertainty. Three scenarios are investigated: no improvement of the systematic uncertainties, a reduction by 20 % and the halving of systematic uncertainties.

Fig. 79 shows the evolution of the significance with an increasing amount of analyzed data. If there were no improvements in the systematic uncertainties, the process $t\bar{t}\gamma$ should be discovered at an integrated luminosity of $\approx 5.7\,\text{fb}^{-1}$. A reduction by 20 % (50 %) would bring this limit down to $\approx 3.9\,\text{fb}^{-1}$ ($\approx 2.9\,\text{fb}^{-1}$) respectively.

Hence, the $5\,\sigma$ discovery boundary cannot be reached with the $5\,\text{fb}^{-1}$ of ATLAS data collected at $\sqrt{s} = 7\,\text{TeV}$ in 2011 if no reduction of the systematic uncertainties can be achieved.

13 RESULTS

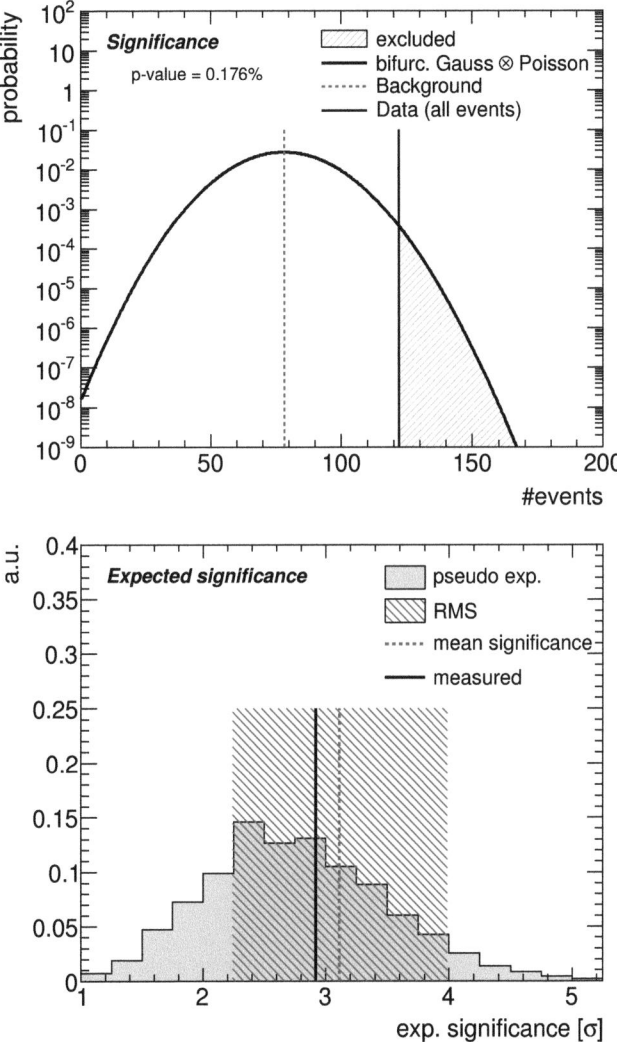

Figure 78: Significance of the measurement, taken from actual data (upper plot) and expected significance estimated from 3000 pseudo experiments (lower plot). The measured and the expected significance are in good agreement.

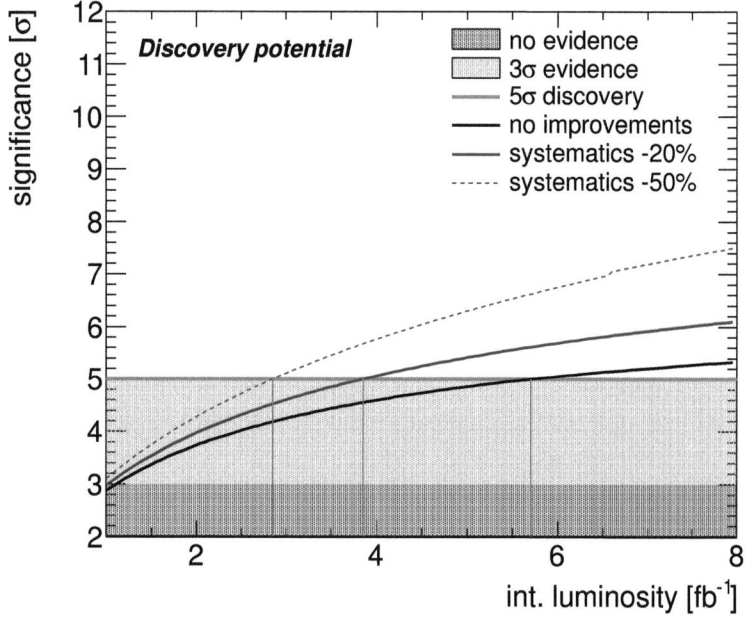

Figure 79: Discovery potential of the process $t\bar{t}\gamma$ depending on the collected amount of data, shown for three different scenarios of improvements of systematic uncertainties: no improvement (solid black), a decrease of the systematic uncertainties by 20% (solid blue) and halving of uncertainties (dashed blue).

14 Summary and Outlook

The cross section of top quark pair production with the additional production of a prompt photon ($t\bar{t}\gamma$) has been measured using $1.04\,\text{fb}^{-1}$ of data recorded by the ATLAS detector in 2011. The measurement is based on a template fit method exploiting different shapes of the track isolation $p_{\text{T,iso}}^{\text{cone20}}$ of photons being produced within jet fragmentation and hadronization (hadron fakes) on the one hand and prompt photons produced within the matrix element of the 7-particle final state process $pp \to b\bar{b}q\bar{q}'\ell\nu\gamma$ (hard process) on the other hand.

The full ME inherently includes top quark pair production and decay accompanied by the emission of an additional photon and is believed to be dominated by top quark propagators internally.

The template fit method yields an expectation of $N_{\text{exp}}(t\bar{t}\gamma) = 1880 \pm 510$ events. Together with all systematic uncertainties, which sum up to $27.2\,\%$ and the uncertainty of the luminosity measurement of $3.7\,\%$, this result translates into a cross section of

$$\sigma_{t\bar{t}\gamma} \times \text{BR} = \left[1.84 \pm 0.49(\text{stat.})^{+0.50}_{-0.50}(\text{syst.}) \pm 0.07(\text{lumi.})\right]\,\text{pb}\,.$$

The result has a significance of $2.9\,\sigma$ w.r.t to the background-only hypothesis. Hence, the hypothesis that at least some amount of $t\bar{t}\gamma$ signal is present in data is almost evident.

Background contributions from various sources have been estimated, using data-driven methods whenever applicable.

The contributions can be distinguished in three classes: hadron fakes, electrons that have been mis-identified as photons ($e \to \gamma$ fakes) and contributions of real prompt photons from processes other than $t\bar{t}\gamma$.

The overall number of hadron fakes has been estimated by a hadron fake isolation template, which covers the contribution of hadron fakes from all possible physical processes.

The contributions of $e \to \gamma$ fakes have been evaluated using the truth information available in MC simulations and are multiplied with a correction factor in order to account for discrepancies of electron mis-identification in data and MC simulation (fake rate SFs). The biggest amount of $e \to \gamma$ fakes stems from di-leptonic $t\bar{t}$ processes and yields 6.7 ± 0.8(stat.) ± 0.5(syst.) events in the e+jets channel and 9.4 ± 1.1(stat.) ± 0.7(syst.) events in the μ+jets channel. Other processes like Z+jets, di-boson and single top quark production give rise to a minor contribution to $e \to \gamma$ fakes as well as remaining $t\bar{t}$ events, where a real photon had been radiated very close to an electron, which cannot be resolved as two distinct objects experimentally. The overall estimation of number of events with an electron mis-identified as a photon yields 8.2 ± 0.6(stat.) ± 0.7(syst.) in the e+jets channel and 10.25 ± 0.04(stat.) ± 0.21(syst.) in the μ+jets channel.

The number of prompt photons produced by background processes has been estimated for each background source separately and has been added to the template fit using the prompt photon template.

The largest background contributions of prompt photons originate from the production of a single W boson with the associated production of jets and the additional emission of a prompt photon (W+jets+γ) and yields 3.5 ± 0.6(stat.) ± 1.0(syst.) events in the e+jets channel and 4.2 ± 0.5(stat.) ± 1.2(syst.) events in the μ+jets channel respectively.

The background contribution of multijet processes with the additional radiation of a prompt photon (QCD+γ) has been determined using data-driven techniques and yields $0.39^{+0.40}_{-0.39} \pm 0.39$ in the e+jets channel and $(0.24^{+0.25}_{-0.24} \pm 0.24)$ in the μ+jets channel.

Other background sources like Z+jets, di-boson and single top quark production have been completely estimated from MC simulation; the overall number of these remaining non-$t\bar{t}$ background events is 2.2 ± 1.1(stat.) ± 0.8(syst.) in the e+jets channel and

$1.8 \pm 1.1(\text{stat.}) \pm 0.6(\text{syst.})$ in the μ+jets channel.

Systematic uncertainties have been evaluated with an ensemble of 3000 pseudo experiments performed on pseudo data which has been created from MC predictions and the expected yield of hadron fakes from the template fit. Each systematic uncertainty has been considered separately and has been applied on the event selection efficiency, the expectation of background events and the isolation templates where applicable, giving rise to an overall number of 35 sources of systematic uncertainties.

The distributions of the photon isolation extracted from pseudo data have been varied according to Poissonian fluctuations and the difference between the averages of the distributions of the number of expected signal events once with and once without applying the systematic variation has been quoted as systematic uncertainty. The combination of all systematic uncertainty yields $^{+27.3}_{-27.0}\,\%$ ($\pm 27.2\,\%$ symmetrized).

The evolution of the significance of the result of the cross section measurement with increasing integrated luminosity has been studied considering three different scenarios of improvement of the systematic uncertainties. With the given magnitude of systematic uncertainties, the $5\,\sigma$ boundary of discovery can be achieved at an integrated luminosity of $\approx 5.7\,\text{fb}^{-1}$ without any improvements the systematic uncertainties. A reduction by $20\,\%$ makes a discovery possible for $\approx 3.9\,\text{fb}^{-1}$ of data, a halving of systematic uncertainties already for $\approx 2.9\,\text{fb}^{-1}$.

Hence, the process $pp \to t\bar{t}\gamma$ should be discovered at a significance of $\geq 5\,\sigma$ using the $5\,\text{fb}^{-1}$ of data collected at $\sqrt{s} = 7\,\text{TeV}$ with slight, further improvements on the systematic uncertainties. This amount of data is planned to be analyzed in the semi-leptonic decay mode by next year. Besides, the di-leptonic final state $pp \to \ell^+\nu_\ell \ell^- \bar{\nu}_\ell b\bar{b}\gamma$ is currently being investigated. The calculation of a fiducial cross section and a corresponding k-factor which adopts all phase space cuts of the $t\bar{t}\gamma$ signal MC simulation is being evaluated.

By the end of September 2012, $\approx 14\,\text{fb}^{-1}$ of data at a CMS energy of 8 TeV had been recorded by the ATLAS detector. The integrated luminosity is hence reaching a regime where differential cross sections can be analyzed in order to study anomalous photon couplings to the top quark.

Part III
Appendix

A Breakdown of Monte-Carlo Contributions

The colored control plots presented in Sec. 10.8 (Fig. 41 to 48) and the corresponding table Tab. 8 show summarized event yields for the Z+jets, W+jets, di-boson and single top contributions. In fact, these numbers a made up from several MC samples.

Tab. 20 and Tab. 21 sum up all single values that contribute to the W+jets event yield for the e+jets and the μ+jets channel separately; Tab. 20 and Tab. 21 show the composition of the Z+jets yield respectively.

The di-boson contributions are presented in Tab. 24 and Tab. 25, those for single top quark production in Tab. 26 and Tab. 27.

Sample	pretag	≥ 1 b-tag
$W \to e\nu + 0p$	262 ± 9	28 ± 5
$W \to e\nu + 1p$	240 ± 60	1 ± 1
$W \to e\nu + 2p$	430 ± 150	35 ± 12
$W \to e\nu + 3p$	1000 ± 400	80 ± 30
$W \to e\nu + 4p$	1400 ± 700	130 ± 60
$W \to e\nu + 5p$	800 ± 400	75 ± 40
$W \to \tau\nu + 0p$	0 ± 0	0 ± 0
$W \to \tau\nu + 1p$	0 ± 0	0 ± 0
$W \to \tau\nu + 2p$	11 ± 4	1 ± 1
$W \to \tau\nu + 3p$	36 ± 15	2 ± 1
$W \to \tau\nu + 4p$	60 ± 30	8 ± 4
$W \to \tau\nu + 5p$	35 ± 20	4 ± 2
$W + b\bar{b} + 0p$	8 ± 4	5 ± 3
$W + b\bar{b} + 1p$	30 ± 20	19 ± 12
$W + b\bar{b} + 2p$	110 ± 70	80 ± 50
$W + b\bar{b} + 3p$	220 ± 160	130 ± 90
$W + c\bar{c} + 0p$	10 ± 6	2 ± 1
$W + c\bar{c} + 1p$	65 ± 40	14 ± 9
$W + c\bar{c} + 2p$	150 ± 100	30 ± 20
$W + c\bar{c} + 3p$	450 ± 320	80 ± 60
$W + c(\bar{c}) + 0p$	40 ± 16	8 ± 3
$W + c(\bar{c}) + 1p$	90 ± 40	20 ± 10
$W + c(\bar{c}) + 2p$	200 ± 100	50 ± 25
$W + c(\bar{c}) + 3p$	40 ± 25	11 ± 6
$W + c(\bar{c}) + 4p$	180 ± 100	45 ± 30

Table 20: Composition of the W+jets contributions relevant for the e+jets channel, shown for the event selection before and after the b-tagging requirement. The uncertainties contain the uncertainties due to Berends-Giele scaling and due to the HF scaling.

A BREAKDOWN OF MONTE-CARLO CONTRIBUTIONS 253

Sample	pretag	≥ 1 b-tag
$W \to \mu\nu + 0p$	720 ± 20	50 ± 6
$W \to \mu\nu + 1p$	460 ± 110	17 ± 5
$W \to \mu\nu + 2p$	850 ± 300	60 ± 20
$W \to \mu\nu + 3p$	1700 ± 700	120 ± 50
$W \to \mu\nu + 4p$	2400 ± 1100	170 ± 80
$W \to \mu\nu + 5p$	1300 ± 700	120 ± 70
$W \to \tau\nu + 0p$	1 ± 1	0 ± 0
$W \to \tau\nu + 1p$	9 ± 3	0 ± 0
$W \to \tau\nu + 2p$	50 ± 15	4 ± 1
$W \to \tau\nu + 3p$	110 ± 50	8 ± 4
$W \to \tau\nu + 4p$	160 ± 75	12 ± 6
$W \to \tau\nu + 5p$	90 ± 50	7 ± 4
$W + b\bar{b} + 0p$	9 ± 5	7 ± 4
$W + b\bar{b} + 1p$	70 ± 40	50 ± 30
$W + b\bar{b} + 2p$	240 ± 160	170 ± 110
$W + b\bar{b} + 3p$	400 ± 280	230 ± 160
$W + c\bar{c} + 0p$	5 ± 3	0 ± 0
$W + c\bar{c} + 1p$	110 ± 70	25 ± 16
$W + c\bar{c} + 2p$	360 ± 240	90 ± 60
$W + c\bar{c} + 3p$	400 ± 280	210 ± 150
$W + c(\bar{c}) + 0p$	70 ± 30	7 ± 3
$W + c(\bar{c}) + 1p$	160 ± 75	37 ± 17
$W + c(\bar{c}) + 2p$	360 ± 190	105 ± 55
$W + c(\bar{c}) + 3p$	80 ± 50	24 ± 14
$W + c(\bar{c}) + 4p$	260 ± 160	70 ± 40

Table 21: Composition of the W+jets contributions relevant for the μ+jets channel, shown for the event selection before and after the b-tagging requirement. The uncertainties contain the uncertainties due to Berends-Giele scaling and due to the HF scaling.

Sample	pretag	≥ 1 b-tag
$Z \to e^+e^- + 0p$	30 ± 1	3 ± 1
$Z \to e^+e^- + 1p$	40 ± 10	1 ± 1
$Z \to e^+e^- + 2p$	100 ± 30	10 ± 3
$Z \to e^+e^- + 3p$	180 ± 80	16 ± 7
$Z \to e^+e^- + 4p$	130 ± 60	17 ± 8
$Z \to e^+e^- + 5p$	70 ± 40	7 ± 4
$Z \to \tau^+\tau^- + 0p$	1 ± 1	0 ± 0
$Z \to \tau^+\tau^- + 1p$	2 ± 1	0 ± 0
$Z \to \tau^+\tau^- + 2p$	9 ± 3	3 ± 1
$Z \to \tau^+\tau^- + 3p$	14 ± 6	2 ± 1
$Z \to \tau^+\tau^- + 4p$	12 ± 6	3 ± 1
$Z \to \tau^+\tau^- + 5p$	7 ± 4	1 ± 1
$Z \to e^+e^- + b\bar{b} + 0p$	3 ± 1	2 ± 1
$Z \to e^+e^- + b\bar{b} + 1p$	8 ± 3	6 ± 2
$Z \to e^+e^- + b\bar{b} + 2p$	16 ± 8	11 ± 5
$Z \to e^+e^- + b\bar{b} + 3p$	21 ± 11	11 ± 6
$Z \to \tau^+\tau^- + b\bar{b} + 0p$	0 ± 0	0 ± 0
$Z \to \tau^+\tau^- + b\bar{b} + 1p$	1 ± 1	1 ± 1
$Z \to \tau^+\tau^- + b\bar{b} + 2p$	1 ± 1	1 ± 1
$Z \to \tau^+\tau^- + b\bar{b} + 3p$	2 ± 1	0 ± 0

Table 22: Composition of the Z+jets contributions relevant for the e+jets channel, shown for the event selection before and after the b-tagging requirement. The uncertainties contain the uncertainties due to Berends-Giele scaling.

A BREAKDOWN OF MONTE-CARLO CONTRIBUTIONS

Sample	pretag	≥ 1 b-tag
$Z \to \mu^+\mu^- + 0p$	45 ± 1	4 ± 1
$Z \to \mu^+\mu^- + 1p$	48 ± 12	3 ± 1
$Z \to \mu^+\mu^- + 2p$	80 ± 30	8 ± 3
$Z \to \mu^+\mu^- + 3p$	140 ± 60	11 ± 5
$Z \to \mu^+\mu^- + 4p$	170 ± 80	14 ± 7
$Z \to \mu^+\mu^- + 5p$	100 ± 50	12 ± 7
$Z \to \tau^+\tau^- + 0p$	8 ± 1	1 ± 1
$Z \to \tau^+\tau^- + 1p$	11 ± 3	1 ± 1
$Z \to \tau^+\tau^- + 2p$	38 ± 13	6 ± 2
$Z \to \tau^+\tau^- + 3p$	90 ± 40	11 ± 5
$Z \to \tau^+\tau^- + 4p$	50 ± 25	14 ± 7
$Z \to \tau^+\tau^- + 5p$	25 ± 14	12 ± 7
$Z \to \mu^+\mu^- + b\bar{b} + 0p$	3 ± 1	2 ± 1
$Z \to \mu^+\mu^- + b\bar{b} + 1p$	4 ± 2	3 ± 1
$Z \to \mu^+\mu^- + b\bar{b} + 2p$	13 ± 6	9 ± 4
$Z \to \mu^+\mu^- + b\bar{b} + 3p$	26 ± 14	16 ± 8
$Z \to \tau^+\tau^- + b\bar{b} + 0p$	0 ± 0	0 ± 0
$Z \to \tau^+\tau^- + b\bar{b} + 1p$	2 ± 1	2 ± 1
$Z \to \tau^+\tau^- + b\bar{b} + 2p$	5 ± 2	4 ± 2
$Z \to \tau^+\tau^- + b\bar{b} + 3p$	7 ± 4	4 ± 2

Table 23: Composition of the Z+jets contributions relevant for the μ+jets channel, shown for the event selection before and after the b-tagging requirement. The uncertainties contain the uncertainties due to Berends-Giele scaling.

Sample	pretag	≥ 1 b-tag
WW	46 ± 3	8 ± 1
WZ	19 ± 1	4 ± 1
ZZ	3 ± 1	1 ± 1

Table 24: Composition of the di-boson contributions in the e+jets channel, shown for the event selection before and after the b-tagging requirement.

Sample	pretag	≥ 1 b-tag
WW	91 ± 5	13 ± 1
WZ	29 ± 2	7 ± 1
ZZ	3 ± 1	1 ± 1

Table 25: Composition of the di-boson contributions in the μ+jets channel, shown for the event selection before and after the b-tagging requirement.

A BREAKDOWN OF MONTE-CARLO CONTRIBUTIONS

Sample	pretag	≥ 1 b-tag
$e\nu$ (s-channel)	8 ± 1	7 ± 1
$\tau\nu$ (s-channel)	0 ± 0	0 ± 0
$e\nu$ (t-channel)	74 ± 3	60 ± 2
$\tau\nu$ (t-channel)	4 ± 1	3 ± 1
Wt	210 ± 15	168 ± 12

Table 26: Composition of the single top contributions in the e+jets channel, shown for the event selection before and after the b-tagging requirement.

Sample	pretag	≥ 1 b-tag
$\mu\nu$ (s-channel)	13 ± 1	11 ± 1
$\tau\nu$ (s-channel)	1 ± 1	1 ± 1
$\mu\nu$ (t-channel)	115 ± 4	94 ± 4
$\tau\nu$ (t-channel)	6 ± 1	5 ± 1
Wt	270 ± 20	211 ± 14

Table 27: Composition of the single top contributions in the μ+jets channel, shown for the event selection before and after the b-tagging requirement.

B List of Monte Carlo Samples

This section lists up all MC samples used in this thesis. Tab. 28 summarizes the $t\bar{t}\gamma$ signal sample as described in Sec. 10.2. Tab. 29 shows details on the baseline MC@NLO $t\bar{t}$ sample. The POWHEG NLO samples (Tab. 31) and the AcerMC LO samples (Tab. 30 and Tab. 33) are used for the evaluation of the systematic uncertainties of the MC prediction of the event selection efficiency (see Sec. 12.3.7).

The W + jets (Z + jets) samples are described in Tab. 34-36 (Tab. 38-40), the corresponding additional HF samples in Tab. 41-43 (Tab. 44-46) respectively.

The W +jets + γ sample needed for the estimation of the W + jets + γ background contribution in Sec. 11.4.2 can be found in Tab. 37. Details on the di-jet sample for the isolation studies presented in Sec. 11.2 are provided in Tab. 32.

The di-boson samples are listed in Tab. 47, those containing single top quark processes in Tab. 48.

B LIST OF MONTE CARLO SAMPLES

Process	$t\bar{t}\gamma \to b\bar{b}j_1j_2\ell\nu\gamma$				
Generator	WHIZARD + HERWIG/JIMMY				
DQ2 sample name	mc10_7TeV.117401.Whizard_Jimmy_TTbarPhoton_SM_NoFullHad				
Specifications	# events	$\sigma \times$ FE [pb]	k-factor	PDF	Transf. package
	498677	0.84	2.33	CTEQ6L1	16.6.5.5.1
Comments	For a detailed description of the sample see Sec. 10.2				

Table 28: $t\bar{t}\gamma$ signal sample.

Process	$t\bar{t} \to b\bar{b}j_1j_2\ell\nu$				
Generator	MC@NLO + HERWIG/JIMMY				
DQ2 sample name	mc10_7TeV.105200.T1_McAtNlo_Jimmy				
Specifications	# events	$\sigma \times$ FE [pb]	k-factor	PDF	Transf. package
	14957047	79.99	1.117	CTEQ66	16.6.5.5.1
Comments	$t(\bar{t})$ decay handled by HERWIG (semi-leptonic decay mode)				

Table 29: Baseline $t\bar{t}$ sample.

Process	$t\bar{t} \to b\bar{b}j_1j_2\ell\nu$				
Generator	AcerMC + Pythia				
DQ2 sample name	mc10_7TeV.105205.AcerMCttbar				
Specifications	# events	$\sigma \times$ FE [pb]	k-factor	PDF	Transf. package
	999328	58.228	1.53	MRST2007lomod	16.6.5.5.1
Comments	$t(\bar{t})$ decay handled by HERWIG (semi-leptonic decay mode)				

Table 30: AcerMC $t\bar{t}$ sample.

Process	$t\bar{t} \to b\bar{b}j_1j_2\ell\nu$				
Generator	POWHEG + Pythia				
DQ2 sample name	mc10_7TeV.105861.TTbar_PowHeg_Pythia				
Specifications	# events	$\sigma \times$ FE [pb]	k-factor	PDF	Transf. package
	2994490	79.17	1.129	MRST2007lomod	16.6.5.5.1
Comments	$t(\bar{t})$ decay handled by Pythia (semi-leptonic decay mode)				
Process	$t\bar{t} \to b\bar{b}j_1j_2\ell\nu$				
Generator	POWHEG + HERWIG				
DQ2 sample name	mc10_7TeV.105860.TTbar_PowHeg_Jimmy				
Specifications	# events	$\sigma \times$ FE [pb]	k-factor	PDF	Transf. package
	2997878	79.17	1.129	MRST2007lomod	16.6.5.5.1
Comments	$t(\bar{t})$ decay handled by HERWIG (semi-leptonic decay mode)				

Table 31: POWHEG $t\bar{t}$ samples.

Process	j_1j_2 (di-jet)				
Generator	Pythia				
DQ2 sample name	mc10_7TeV.105802.JF17_pythia_jet_filter				
Specifications	# events	$\sigma \times$ FE [pb]	k-factor	PDF	Transf. package
	39772064	$84.98 \cdot 10^6$	1.0	MRST2007lomod	16.6.5.5.1
Comments	—				

Table 32: JF17 di-jet sample.

B LIST OF MONTE CARLO SAMPLES

Process	$t\bar{t} \to b\bar{b}j_1j_2\ell\nu$ (ISR varied downwards)				
Generator	AcerMC + Pythia				
DQ2 sample name	mc10_7TeV.117255.AcerMCttbar_isr_down				
Specifications	# events	$\sigma \times$ FE [pb]	k-factor	PDF	Transf. package
	994382	58.23	1.53	MRST2007lomod	16.6.5.5.1
Comments	$t(\bar{t})$ decay handled by Pythia (semi-leptonic decay mode)				

Process	$t\bar{t} \to b\bar{b}j_1j_2\ell\nu$ (ISR varied upwards)				
Generator	AcerMC + Pythia				
DQ2 sample name	mc10_7TeV.117256.AcerMCttbar_isr_up				
Specifications	# events	$\sigma \times$ FE [pb]	k-factor	PDF	Transf. package
	999206	58.23	1.53	MRST2007lomod	16.6.5.5.1
Comments	$t(\bar{t})$ decay handled by Pythia (semi-leptonic decay mode)				

Process	$t\bar{t} \to b\bar{b}j_1j_2\ell\nu$ (FSR varied downwards)				
Generator	AcerMC + Pythia				
DQ2 sample name	mc10_7TeV.117257.AcerMCttbar_fsr_down				
Specifications	# events	$\sigma \times$ FE [pb]	k-factor	PDF	Transf. package
	999328	58.23	1.53	MRST2007lomod	16.6.5.5.1
Comments	$t(\bar{t})$ decay handled by Pythia (semi-leptonic decay mode)				

Process	$t\bar{t} \to b\bar{b}j_1j_2\ell\nu$ (FSR varied upwards)				
Generator	AcerMC + Pythia				
DQ2 sample name	mc10_7TeV.117258.AcerMCttbar_fsr_up				
Specifications	# events	$\sigma \times$ FE [pb]	k-factor	PDF	Transf. package
	997758	58.23	1.53	MRST2007lomod	16.6.5.5.1
Comments	$t(\bar{t})$ decay handled by Pythia (semi-leptonic decay mode)				

Process	$t\bar{t} \to b\bar{b}j_1j_2\ell\nu$ (ISR+FSR varied downwards)				
Generator	AcerMC + Pythia				
DQ2 sample name	mc10_7TeV.117259.AcerMCttbar_isr_down_fsr_down				
Specifications	# events	$\sigma \times$ FE [pb]	k-factor	PDF	Transf. package
	994382	58.23	1.53	MRST2007lomod	16.6.5.5.1
Comments	$t(\bar{t})$ decay handled by Pythia (semi-leptonic decay mode)				

Process	$t\bar{t} \to b\bar{b}j_1j_2\ell\nu$ (ISR+FSR varied upwards)				
Generator	AcerMC + Pythia				
DQ2 sample name	mc10_7TeV.117260.AcerMCttbar_isr_up_fsr_up				
Specifications	# events	$\sigma \times$ FE [pb]	k-factor	PDF	Transf. package
	994382	58.23	1.53	MRST2007lomod	16.6.5.5.1
Comments	$t(\bar{t})$ decay handled by Pythia (semi-leptonic decay mode)				

Table 33: AcerMC $t\bar{t}$ samples with ISR/FSR variations.

Process	$W \to e\nu + 0p$				
Generator	ALPGEN + HERWIG/JIMMY				
DQ2 sample name	mc10_7TeV.107680.AlpgenJimmyWenuNp0_pt20				
Specifications	# events	$\sigma \times$ FE [pb]	k-factor	PDF	Transf. package
	3455037	6921.60	1.20	CTEQ6L1	16.6.5.5.1
Comments	MLM matching for partons with $p_T > 20\,\mathrm{GeV}$				

Process	$W \to e\nu + 1p$				
Generator	ALPGEN + HERWIG/JIMMY				
DQ2 sample name	mc10_7TeV.107681.AlpgenJimmyWenuNp1_pt20				
Specifications	# events	$\sigma \times$ FE [pb]	k-factor	PDF	Transf. package
	641361	1304.30	1.20	CTEQ6L1	16.6.5.5.1
Comments	MLM matching for partons with $p_T > 20\,\mathrm{GeV}$				

Process	$W \to e\nu + 2p$				
Generator	ALPGEN + HERWIG/JIMMY				
DQ2 sample name	mc10_7TeV.107682.AlpgenJimmyWenuNp2_pt20				
Specifications	# events	$\sigma \times$ FE [pb]	k-factor	PDF	Transf. package
	3768265	378.29	1.20	CTEQ6L1	16.6.5.5.1
Comments	MLM matching for partons with $p_T > 20\,\mathrm{GeV}$				

Process	$W \to e\nu + 3p$				
Generator	ALPGEN + HERWIG/JIMMY				
DQ2 sample name	mc10_7TeV.107683.AlpgenJimmyWenuNp3_pt20				
Specifications	# events	$\sigma \times$ FE [pb]	k-factor	PDF	Transf. package
	1009641	101.43	1.20	CTEQ6L1	16.6.5.5.1
Comments	MLM matching for partons with $p_T > 20\,\mathrm{GeV}$				

Process	$W \to e\nu + 4p$				
Generator	ALPGEN + HERWIG/JIMMY				
DQ2 sample name	mc10_7TeV.107684.AlpgenJimmyWenuNp4_pt20				
Specifications	# events	$\sigma \times$ FE [pb]	k-factor	PDF	Transf. package
	249869	25.87	1.20	CTEQ6L1	16.6.5.5.1
Comments	MLM matching for partons with $p_T > 20\,\mathrm{GeV}$				

Process	$W \to e\nu + 5p$				
Generator	ALPGEN + HERWIG/JIMMY				
DQ2 sample name	mc10_7TeV.107685.AlpgenJimmyWenuNp5_pt20				
Specifications	# events	$\sigma \times$ FE [pb]	k-factor	PDF	Transf. package
	69953	7.00	1.20	CTEQ6L1	16.6.5.5.1
Comments	MLM matching for partons with $p_T > 20\,\mathrm{GeV}$				

Table 34: $W \to e\nu$ + jets $(0\ldots 5$ additional partons$)$.

B LIST OF MONTE CARLO SAMPLES

Process	$W \to \mu\nu + 0p$				
Generator	ALPGEN + HERWIG/JIMMY				
DQ2 sample name	mc10_7TeV.107690.AlpgenJimmyWmununNp0_pt20				
Specifications	# events	$\sigma \times$ FE [pb]	k-factor	PDF	Transf. package
	3466523	6919.60	1.20	CTEQ6L1	16.6.5.5.1
Comments	MLM matching for partons with $p_T > 20\,\text{GeV}$				

Process	$W \to \mu\nu + 1p$				
Generator	ALPGEN + HERWIG/JIMMY				
DQ2 sample name	mc10_7TeV.107691.AlpgenJimmyWmununNp1_pt20				
Specifications	# events	$\sigma \times$ FE [pb]	k-factor	PDF	Transf. package
	641867	1304.20	1.20	CTEQ6L1	16.6.5.5.1
Comments	MLM matching for partons with $p_T > 20\,\text{GeV}$				

Process	$W \to \mu\nu + 2p$				
Generator	ALPGEN + HERWIG/JIMMY				
DQ2 sample name	mc10_7TeV.107692.AlpgenJimmyWmununNp2_pt20				
Specifications	# events	$\sigma \times$ FE [pb]	k-factor	PDF	Transf. package
	3768893	377.83	1.20	CTEQ6L1	16.6.5.5.1
Comments	MLM matching for partons with $p_T > 20\,\text{GeV}$				

Process	$W \to \mu\nu + 3p$				
Generator	ALPGEN + HERWIG/JIMMY				
DQ2 sample name	mc10_7TeV.107693.AlpgenJimmyWmununNp3_pt20				
Specifications	# events	$\sigma \times$ FE [pb]	k-factor	PDF	Transf. package
	1009589	101.88	1.20	CTEQ6L1	16.6.5.5.1
Comments	MLM matching for partons with $p_T > 20\,\text{GeV}$				

Process	$W \to \mu\nu + 4p$				
Generator	ALPGEN + HERWIG/JIMMY				
DQ2 sample name	mc10_7TeV.107694.AlpgenJimmyWmununNp4_pt20				
Specifications	# events	$\sigma \times$ FE [pb]	k-factor	PDF	Transf. package
	254879	25.75	1.20	CTEQ6L1	16.6.5.5.1
Comments	MLM matching for partons with $p_T > 20\,\text{GeV}$				

Process	$W \to \mu\nu + 5p$				
Generator	ALPGEN + HERWIG/JIMMY				
DQ2 sample name	mc10_7TeV.107695.AlpgenJimmyWmununNp5_pt20				
Specifications	# events	$\sigma \times$ FE [pb]	k-factor	PDF	Transf. package
	69958	6.92	1.20	CTEQ6L1	16.6.5.5.1
Comments	MLM matching for partons with $p_T > 20\,\text{GeV}$				

Table 35: $W \to \mu\nu$ + jets (0...5 additional partons).

Process	$W \to \tau\nu + 0p$				
Generator	ALPGEN + HERWIG/JIMMY				
DQ2 sample name	mc10_7TeV.107700.AlpgenJimmyWtaunuNp0_pt20				
Specifications	# events	$\sigma \times$ FE [pb]	k-factor	PDF	Transf. package
	3416438	6918.60	1.20	CTEQ6L1	16.6.5.5.1
Comments	MLM matching for partons with $p_T > 20\,\text{GeV}$				

Process	$W \to \tau\nu + 1p$				
Generator	ALPGEN + HERWIG/JIMMY				
DQ2 sample name	mc10_7TeV.107701.AlpgenJimmyWtaunuNp1_pt20				
Specifications	# events	$\sigma \times$ FE [pb]	k-factor	PDF	Transf. package
	641809	1303.20	1.20	CTEQ6L1	16.6.5.5.1
Comments	MLM matching for partons with $p_T > 20\,\text{GeV}$				

Process	$W \to \tau\nu + 2p$				
Generator	ALPGEN + HERWIG/JIMMY				
DQ2 sample name	mc10_7TeV.107702.AlpgenJimmyWtaunuNp2_pt20				
Specifications	# events	$\sigma \times$ FE [pb]	k-factor	PDF	Transf. package
	3768750	378.18	1.20	CTEQ6L1	16.6.5.5.1
Comments	MLM matching for partons with $p_T > 20\,\text{GeV}$				

Process	$W \to \tau\nu + 3p$				
Generator	ALPGEN + HERWIG/JIMMY				
DQ2 sample name	mc10_7TeV.107703.AlpgenJimmyWtaunuNp3_pt20				
Specifications	# events	$\sigma \times$ FE [pb]	k-factor	PDF	Transf. package
	1009548	101.51	1.20	CTEQ6L1	16.6.5.5.1
Comments	MLM matching for partons with $p_T > 20\,\text{GeV}$				

Process	$W \to \tau\nu + 4p$				
Generator	ALPGEN + HERWIG/JIMMY				
DQ2 sample name	mc10_7TeV.107704.AlpgenJimmyWtaunuNp4_pt20				
Specifications	# events	$\sigma \times$ FE [pb]	k-factor	PDF	Transf. package
	249853	25.64	1.20	CTEQ6L1	16.6.5.5.1
Comments	MLM matching for partons with $p_T > 20\,\text{GeV}$				

Process	$W \to \tau\nu + 5p$				
Generator	ALPGEN + HERWIG/JIMMY				
DQ2 sample name	mc10_7TeV.107705.AlpgenJimmyWtaunuNp5_pt20				
Specifications	# events	$\sigma \times$ FE [pb]	k-factor	PDF	Transf. package
	63692	7.04	1.20	CTEQ6L1	16.6.5.5.1
Comments	MLM matching for partons with $p_T > 20\,\text{GeV}$				

Table 36: $W \to \tau\nu$ + jets ($0 \ldots 5$ additional partons).

B LIST OF MONTE CARLO SAMPLES

Process	$W + \gamma + 0p$				
Generator	ALPGEN + HERWIG				
DQ2 sample name	mc10_7TeV.117410.AlpgenJimmyWgammaNp0_pt20				
Specifications	# events	$\sigma \times$ FE [pb]	k-factor	PDF	Transf. package
	1409784	213.06	1.00	CTEQ6L1	16.6.5.5.1
Comments	—				

Process	$W + \gamma + 1p$				
Generator	ALPGEN + HERWIG				
DQ2 sample name	mc10_7TeV.117411.AlpgenJimmyWgammaNp1_pt20				
Specifications	# events	$\sigma \times$ FE [pb]	k-factor	PDF	Transf. package
	499887	52.20	1.00	CTEQ6L1	16.6.5.5.1
Comments	—				

Process	$W + \gamma + 2p$				
Generator	ALPGEN + HERWIG				
DQ2 sample name	mc10_7TeV.117412.AlpgenJimmyWgammaNp2_pt20				
Specifications	# events	$\sigma \times$ FE [pb]	k-factor	PDF	Transf. package
	174939	17.22	1.00	CTEQ6L1	16.6.5.5.1
Comments	—				

Process	$W + \gamma + 3p$				
Generator	ALPGEN + HERWIG				
DQ2 sample name	mc10_7TeV.117413.AlpgenJimmyWgammaNp3_pt20				
Specifications	# events	$\sigma \times$ FE [pb]	k-factor	PDF	Transf. package
	264886	5.34	1.00	CTEQ6L1	16.6.5.5.1
Comments	—				

Process	$W + \gamma + 4p$				
Generator	ALPGEN + HERWIG				
DQ2 sample name	mc10_7TeV.117414.AlpgenJimmyWgammaNp4_pt20				
Specifications	# events	$\sigma \times$ FE [pb]	k-factor	PDF	Transf. package
	69961	1.38	1.00	CTEQ6L1	16.6.5.5.1
Comments	—				

Process	$W + \gamma + 5p$				
Generator	ALPGEN + HERWIG				
DQ2 sample name	mc10_7TeV.117415.AlpgenJimmyWgammaNp5_pt20				
Specifications	# events	$\sigma \times$ FE [pb]	k-factor	PDF	Transf. package
	19979	0.34	1.00	CTEQ6L1	16.6.5.5.1
Comments	—				

Table 37: W + jets + γ samples (0 ... 5 additional partons).

Process	$Z \to e^+e^- + 0p$				
Generator	ALPGEN + HERWIG/JIMMY				
DQ2 sample name	mc10_7TeV.107650.AlpgenJimmyZeeNp0_pt20				
Specifications	# events	$\sigma \times$ FE [pb]	k-factor	PDF	Transf. package
	6612265	668.32	1.25	CTEQ6L1	16.6.5.5.1
Comments	MLM matching for partons with $p_T > 20\,\text{GeV}$				

Process	$Z \to e^+e^- + 1p$				
Generator	ALPGEN + HERWIG/JIMMY				
DQ2 sample name	mc10_7TeV.107651.AlpgenJimmyZeeNp1_pt20				
Specifications	# events	$\sigma \times$ FE [pb]	k-factor	PDF	Transf. package
	1333745	134.36	1.25	CTEQ6L1	16.6.5.5.1
Comments	MLM matching for partons with $p_T > 20\,\text{GeV}$				

Process	$Z \to e^+e^- + 2p$				
Generator	ALPGEN + HERWIG/JIMMY				
DQ2 sample name	mc10_7TeV.107652.AlpgenJimmyZeeNp2_pt20				
Specifications	# events	$\sigma \times$ FE [pb]	k-factor	PDF	Transf. package
	404873	40.54	1.25	CTEQ6L1	16.6.5.5.1
Comments	MLM matching for partons with $p_T > 20\,\text{GeV}$				

Process	$Z \to e^+e^- + 3p$				
Generator	ALPGEN + HERWIG/JIMMY				
DQ2 sample name	mc10_7TeV.107653.AlpgenJimmyZeeNp3_pt20				
Specifications	# events	$\sigma \times$ FE [pb]	k-factor	PDF	Transf. package
	109942	11.16	1.25	CTEQ6L1	16.6.5.5.1
Comments	MLM matching for partons with $p_T > 20\,\text{GeV}$				

Process	$Z \to e^+e^- + 4p$				
Generator	ALPGEN + HERWIG/JIMMY				
DQ2 sample name	mc10_7TeV.107654.AlpgenJimmyZeeNp4_pt20				
Specifications	# events	$\sigma \times$ FE [pb]	k-factor	PDF	Transf. package
	29992	2.88	1.25	CTEQ6L1	16.6.5.5.1
Comments	MLM matching for partons with $p_T > 20\,\text{GeV}$				

Process	$Z \to e^+e^- + 5p$				
Generator	ALPGEN + HERWIG/JIMMY				
DQ2 sample name	mc10_7TeV.107655.AlpgenJimmyZeeNp5_pt20				
Specifications	# events	$\sigma \times$ FE [pb]	k-factor	PDF	Transf. package
	8992	0.83	1.25	CTEQ6L1	16.6.5.5.1
Comments	MLM matching for partons with $p_T > 20\,\text{GeV}$				

Table 38: $Z \to e^+e^-$ + jets ($0\ldots 5$ additional partons).

B LIST OF MONTE CARLO SAMPLES

Process	$Z \to \mu^+\mu^- + 0p$				
Generator	ALPGEN + HERWIG/JIMMY				
DQ2 sample name	mc10_7TeV.107660.AlpgenJimmyZmumuNp0_pt20				
Specifications	# events	$\sigma \times$ FE [pb]	k-factor	PDF	Transf. package
	6619010	668.68	1.25	CTEQ6L1	16.6.5.5.1
Comments	MLM matching for partons with $p_T > 20\,\text{GeV}$				
Process	$Z \to \mu^+\mu^- + 1p$				
Generator	ALPGEN + HERWIG/JIMMY				
DQ2 sample name	mc10_7TeV.107661.AlpgenJimmyZmumuNp1_pt20				
Specifications	# events	$\sigma \times$ FE [pb]	k-factor	PDF	Transf. package
	1334723	134.14	1.25	CTEQ6L1	16.6.5.5.1
Comments	MLM matching for partons with $p_T > 20\,\text{GeV}$				
Process	$Z \to \mu^+\mu^- + 2p$				
Generator	ALPGEN + HERWIG/JIMMY				
DQ2 sample name	mc10_7TeV.107662.AlpgenJimmyZmumuNp2_pt20				
Specifications	# events	$\sigma \times$ FE [pb]	k-factor	PDF	Transf. package
	403886	40.33	1.25	CTEQ6L1	16.6.5.5.1
Comments	MLM matching for partons with $p_T > 20\,\text{GeV}$				
Process	$Z \to \mu^+\mu^- + 3p$				
Generator	ALPGEN + HERWIG/JIMMY				
DQ2 sample name	mc10_7TeV.107663.AlpgenJimmyZmumuNp3_pt20				
Specifications	# events	$\sigma \times$ FE [pb]	k-factor	PDF	Transf. package
	109954	11.19	1.25	CTEQ6L1	16.6.5.5.1
Comments	MLM matching for partons with $p_T > 20\,\text{GeV}$				
Process	$Z \to \mu^+\mu^- + 4p$				
Generator	ALPGEN + HERWIG/JIMMY				
DQ2 sample name	mc10_7TeV.107664.AlpgenJimmyZmumuNp4_pt20				
Specifications	# events	$\sigma \times$ FE [pb]	k-factor	PDF	Transf. package
	29978	2.75	1.25	CTEQ6L1	16.6.5.5.1
Comments	MLM matching for partons with $p_T > 20\,\text{GeV}$				
Process	$Z \to \mu^+\mu^- + 5p$				
Generator	ALPGEN + HERWIG/JIMMY				
DQ2 sample name	mc10_7TeV.107665.AlpgenJimmyZmumuNp5_pt20				
Specifications	# events	$\sigma \times$ FE [pb]	k-factor	PDF	Transf. package
	9993	0.77	1.25	CTEQ6L1	16.6.5.5.1
Comments	MLM matching for partons with $p_T > 20\,\text{GeV}$				

Table 39: $Z \to \mu^+\mu^-$ + jets (0...5 additional partons).

Process	$Z \to \tau^+\tau^- + 0p$				
Generator	ALPGEN + HERWIG/JIMMY				
DQ2 sample name	mc10_7TeV.107670.AlpgenJimmyZtautauNp0_pt20				
Specifications	# events	$\sigma \times$ FE [pb]	k-factor	PDF	Transf. package
	6618801	668.40	1.25	CTEQ6L1	16.6.5.5.1
Comments	MLM matching for partons with $p_T > 20\,\mathrm{GeV}$				

Process	$Z \to \tau^+\tau^- + 1p$				
Generator	ALPGEN + HERWIG/JIMMY				
DQ2 sample name	mc10_7TeV.107671.AlpgenJimmyZtautauNp1_pt20				
Specifications	# events	$\sigma \times$ FE [pb]	k-factor	PDF	Transf. package
	1334664	134.81	1.25	CTEQ6L1	16.6.5.5.1
Comments	MLM matching for partons with $p_T > 20\,\mathrm{GeV}$				

Process	$Z \to \tau^+\tau^- + 2p$				
Generator	ALPGEN + HERWIG/JIMMY				
DQ2 sample name	mc10_7TeV.107672.AlpgenJimmyZtautauNp2_pt20				
Specifications	# events	$\sigma \times$ FE [pb]	k-factor	PDF	Transf. package
	404853	40.36	1.25	CTEQ6L1	16.6.5.5.1
Comments	MLM matching for partons with $p_T > 20\,\mathrm{GeV}$				

Process	$Z \to \tau^+\tau^- + 3p$				
Generator	ALPGEN + HERWIG/JIMMY				
DQ2 sample name	mc10_7TeV.107673.AlpgenJimmyZtautauNp3_pt20				
Specifications	# events	$\sigma \times$ FE [pb]	k-factor	PDF	Transf. package
	109944	11.25	1.25	CTEQ6L1	16.6.5.5.1
Comments	MLM matching for partons with $p_T > 20\,\mathrm{GeV}$				

Process	$Z \to \tau^+\tau^- + 4p$				
Generator	ALPGEN + HERWIG/JIMMY				
DQ2 sample name	mc10_7TeV.107674.AlpgenJimmyZtautauNp4_pt20				
Specifications	# events	$\sigma \times$ FE [pb]	k-factor	PDF	Transf. package
	29982	2.79	1.25	CTEQ6L1	16.6.5.5.1
Comments	MLM matching for partons with $p_T > 20\,\mathrm{GeV}$				

Process	$Z \to \tau^+\tau^- + 5p$				
Generator	ALPGEN + HERWIG/JIMMY				
DQ2 sample name	mc10_7TeV.107675.AlpgenJimmyZtautauNp5_pt20				
Specifications	# events	$\sigma \times$ FE [pb]	k-factor	PDF	Transf. package
	9993	0.77	1.25	CTEQ6L1	16.6.5.5.1
Comments	MLM matching for partons with $p_T > 20\,\mathrm{GeV}$				

Table 40: $Z \to \tau^+\tau^-$ + jets $(0\ldots 5$ additional partons).

B LIST OF MONTE CARLO SAMPLES

Process	$W + b\bar{b} + 0p$				
Generator	ALPGEN + HERWIG/JIMMY				
DQ2 sample name	mc10_7TeV.107280.AlpgenJimmyWbbFullNp0_pt20				
Specifications	# events	$\sigma \times$ FE [pb]	k-factor	PDF	Transf. package
	474933	47.32	1.20	CTEQ6L1	16.6.5.5.1
Comments	$W \to \ell\nu$ ($\ell = e, \mu, \tau$) decay handled by HERWIG				
	no phase space cuts on heavy quarks				
	MLM matching for partons with $p_T > 20\,\text{GeV}$				

Process	$W + b\bar{b} + 1p$				
Generator	ALPGEN + HERWIG/JIMMY				
DQ2 sample name	mc10_7TeV.107281.AlpgenJimmyWbbFullNp1_pt20				
Specifications	# events	$\sigma \times$ FE [pb]	k-factor	PDF	Transf. package
	204933	35.77	1.20	CTEQ6L1	16.6.5.5.1
Comments	$W \to \ell\nu$ ($\ell = e, \mu, \tau$) decay handled by HERWIG				
	no phase space cuts on heavy quarks				
	MLM matching for partons with $p_T > 20\,\text{GeV}$				

Process	$W + b\bar{b} + 2p$				
Generator	ALPGEN + HERWIG/JIMMY				
DQ2 sample name	mc10_7TeV.107282.AlpgenJimmyWbbFullNp2_pt20				
Specifications	# events	$\sigma \times$ FE [pb]	k-factor	PDF	Transf. package
	174942	17.34	1.20	CTEQ6L1	16.6.5.5.1
Comments	$W \to \ell\nu$ ($\ell = e, \mu, \tau$) decay handled by HERWIG				
	no phase space cuts on heavy quarks				
	MLM matching for partons with $p_T > 20\,\text{GeV}$				

Process	$W + b\bar{b} + 3p$				
Generator	ALPGEN + HERWIG/JIMMY				
DQ2 sample name	mc10_7TeV.107283.AlpgenJimmyWbbFullNp3_pt20				
Specifications	# events	$\sigma \times$ FE [pb]	k-factor	PDF	Transf. package
	69969	6.63	1.20	CTEQ6L1	16.6.5.5.1
Comments	$W \to \ell\nu$ ($\ell = e, \mu, \tau$) decay handled by HERWIG				
	no phase space cuts on heavy quarks				
	MLM matching for partons with $p_T > 20\,\text{GeV}$				

Table 41: $W + b\bar{b}$ + jets (0 ... 3 additional partons).

Process	$W + c\bar{c} + 0p$				
Generator	ALPGEN + HERWIG/JIMMY				
DQ2 sample name	mc10_7TeV.117284.AlpgenWccFullNp0_pt20				
Specifications	# events	$\sigma \times FE$ [pb]	k-factor	PDF	Transf. package
	254955	127.53	1.20	CTEQ6L1	16.6.5.5.1
Comments	$W \to \ell\nu$ ($\ell = e, \mu, \tau$) decay handled by HERWIG				
	no phase space cuts on heavy quarks				
	MLM matching for partons with $p_T > 20$ GeV				
Process	$W + c\bar{c} + 1p$				
Generator	ALPGEN + HERWIG/JIMMY				
DQ2 sample name	mc10_7TeV.117285.AlpgenWccFullNp1_pt20				
Specifications	# events	$\sigma \times FE$ [pb]	k-factor	PDF	Transf. package
	206446	104.68	1.20	CTEQ6L1	16.6.5.5.1
Comments	$W \to \ell\nu$ ($\ell = e, \mu, \tau$) decay handled by HERWIG				
	no phase space cuts on heavy quarks				
	MLM matching for partons with $p_T > 20$ GeV				
Process	$W + c\bar{c} + 2p$				
Generator	ALPGEN + HERWIG/JIMMY				
DQ2 sample name	mc10_7TeV.117286.AlpgenWccFullNp2_pt20				
Specifications	# events	$\sigma \times FE$ [pb]	k-factor	PDF	Transf. package
	103464	52.08	1.20	CTEQ6L1	16.6.5.5.1
Comments	$W \to \ell\nu$ ($\ell = e, \mu, \tau$) decay handled by HERWIG				
	no phase space cuts on heavy quarks				
	MLM matching for partons with $p_T > 20$ GeV				
Process	$W + c\bar{c} + 3p$				
Generator	ALPGEN + HERWIG/JIMMY				
DQ2 sample name	mc10_7TeV.117287.AlpgenWccFullNp3_pt20				
Specifications	# events	$\sigma \times FE$ [pb]	k-factor	PDF	Transf. package
	33984	16.96	1.20	CTEQ6L1	16.6.5.5.1
Comments	$W \to \ell\nu$ ($\ell = e, \mu, \tau$) decay handled by HERWIG				
	no phase space cuts on heavy quarks				
	MLM matching for partons with $p_T > 20$ GeV				

Table 42: $W + c\bar{c}$ + jets (0...3 additional partons).

B LIST OF MONTE CARLO SAMPLES

Process	$W + c(\bar{c}) + 0p$				
Generator	ALPGEN + HERWIG/JIMMY				
DQ2 sample name	mc10_7TeV.117293.AlpgenWcNp0_pt20				
Specifications	# events	$\sigma \times$ FE [pb]	k-factor	PDF	Transf. package
	6483825	644.4	1.20	CTEQ6L1	16.6.5.5.1
Comments	$W \to \ell\nu$ ($\ell = e, \mu, \tau$) decay handled by HERWIG $p_T(c(\bar{c})) > 10\,\text{GeV}$, $\Delta R(c(\bar{c}), \text{jet}) > 0.7$ MLM matching for partons with $p_T > 20\,\text{GeV}$				

Process	$W + c(\bar{c}) + 1p$				
Generator	ALPGEN + HERWIG/JIMMY				
DQ2 sample name	mc10_7TeV.117294.AlpgenWcNp1_pt20				
Specifications	# events	$\sigma \times$ FE [pb]	k-factor	PDF	Transf. package
	2069456	205.0	1.20	CTEQ6L1	16.6.5.5.1
Comments	$W \to \ell\nu$ ($\ell = e, \mu, \tau$) decay handled by HERWIG $p_T(c(\bar{c})) > 10\,\text{GeV}$, $\Delta R(c(\bar{c}), \text{jet}) > 0.7$ MLM matching for partons with $p_T > 20\,\text{GeV}$				

Process	$W + c(\bar{c}) + 2p$				
Generator	ALPGEN + HERWIG/JIMMY				
DQ2 sample name	mc10_7TeV.117295.AlpgenWcNp2_pt20				
Specifications	# events	$\sigma \times$ FE [pb]	k-factor	PDF	Transf. package
	517833	50.8	1.20	CTEQ6L1	16.6.5.5.1
Comments	$W \to \ell\nu$ ($\ell = e, \mu, \tau$) decay handled by HERWIG $p_T(c(\bar{c})) > 10\,\text{GeV}$, $\Delta R(c(\bar{c}), \text{jet}) > 0.7$ MLM matching for partons with $p_T > 20\,\text{GeV}$				

Process	$W + c(\bar{c}) + 3p$				
Generator	ALPGEN + HERWIG/JIMMY				
DQ2 sample name	mc10_7TeV.117296.AlpgenWcNp3_pt20				
Specifications	# events	$\sigma \times$ FE [pb]	k-factor	PDF	Transf. package
	114936	11.4	1.20	CTEQ6L1	16.6.5.5.1
Comments	$W \to \ell\nu$ ($\ell = e, \mu, \tau$) decay handled by HERWIG $p_T(c(\bar{c})) > 10\,\text{GeV}$, $\Delta R(c(\bar{c}), \text{jet}) > 0.7$ MLM matching for partons with $p_T > 20\,\text{GeV}$				

Process	$W + c(\bar{c}) + 4p$				
Generator	ALPGEN + HERWIG/JIMMY				
DQ2 sample name	mc10_7TeV.117297.AlpgenWcNp4_pt20				
Specifications	# events	$\sigma \times$ FE [pb]	k-factor	PDF	Transf. package
	29977	2.8	1.20	CTEQ6L1	16.6.5.5.1
Comments	$W \to \ell\nu$ ($\ell = e, \mu, \tau$) decay handled by HERWIG $p_T(c(\bar{c})) > 10\,\text{GeV}$, $\Delta R(c(\bar{c}), \text{jet}) > 0.7$ MLM matching for partons with $p_T > 20\,\text{GeV}$				

Table 43: $W + c(\bar{c}) +$ jets ($0 \ldots 4$ additional partons).

Process	$Z \to e^+e^- + b\bar{b} + 0p$				
Generator	ALPGEN + HERWIG/JIMMY				
DQ2 sample name	mc10_7TeV.109300.AlpgenJimmyZeebbNp0_nofilter				
Specifications	# events	$\sigma \times$ FE [pb]	k-factor	PDF	Transf. package
	149971	6.57	1.25	CTEQ6L1	16.6.5.5.1
Comments	MLM matching for partons with $p_T > 20$ GeV				
Process	$Z \to e^+e^- + b\bar{b} + 1p$				
Generator	ALPGEN + HERWIG/JIMMY				
DQ2 sample name	mc10_7TeV.109301.AlpgenJimmyZeebbNp1_nofilter				
Specifications	# events	$\sigma \times$ FE [pb]	k-factor	PDF	Transf. package
	99977	2.48	1.25	CTEQ6L1	16.6.5.5.1
Comments	MLM matching for partons with $p_T > 20$ GeV				
Process	$Z \to e^+e^- + b\bar{b} + 2p$				
Generator	ALPGEN + HERWIG/JIMMY				
DQ2 sample name	mc10_7TeV.109302.AlpgenJimmyZeebbNp2_nofilter				
Specifications	# events	$\sigma \times$ FE [pb]	k-factor	PDF	Transf. package
	38985	0.89	1.25	CTEQ6L1	16.6.5.5.1
Comments	MLM matching for partons with $p_T > 20$ GeV				
Process	$Z \to e^+e^- + b\bar{b} + 3p$				
Generator	ALPGEN + HERWIG/JIMMY				
DQ2 sample name	mc10_7TeV.109303.AlpgenJimmyZeebbNp3_nofilter				
Specifications	# events	$\sigma \times$ FE [pb]	k-factor	PDF	Transf. package
	9990	0.39	1.25	CTEQ6L1	16.6.5.5.1
Comments	MLM matching for partons with $p_T > 20$ GeV				

Table 44: $Z \to e^+e^- + b\bar{b}$ + jets (0 ... 3 additional partons).

B LIST OF MONTE CARLO SAMPLES

Process	$Z \to \mu^+\mu^- + b\bar{b} + 0p$				
Generator	ALPGEN + HERWIG/JIMMY				
DQ2 sample name	mc10_7TeV.109305.AlpgenJimmyZmumubbNp0_nofilter				
Specifications	# events	$\sigma \times$ FE [pb]	k-factor	PDF	Transf. package
	149971	6.56	1.25	CTEQ6L1	16.6.5.5.1
Comments	MLM matching for partons with $p_T > 20$ GeV				

Process	$Z \to \mu^+\mu^- + b\bar{b} + 1p$				
Generator	ALPGEN + HERWIG/JIMMY				
DQ2 sample name	mc10_7TeV.109306.AlpgenJimmyZmumubbNp1_nofilter				
Specifications	# events	$\sigma \times$ FE [pb]	k-factor	PDF	Transf. package
	99967	2.47	1.25	CTEQ6L1	16.6.5.5.1
Comments	MLM matching for partons with $p_T > 20$ GeV				

Process	$Z \to \mu^+\mu^- + b\bar{b} + 2p$				
Generator	ALPGEN + HERWIG/JIMMY				
DQ2 sample name	mc10_7TeV.109307.AlpgenJimmyZmumubbNp2_nofilter				
Specifications	# events	$\sigma \times$ FE [pb]	k-factor	PDF	Transf. package
	39980	0.89	1.25	CTEQ6L1	16.6.5.5.1
Comments	MLM matching for partons with $p_T > 20$ GeV				

Process	$Z \to \mu^+\mu^- + b\bar{b} + 3p$				
Generator	ALPGEN + HERWIG/JIMMY				
DQ2 sample name	mc10_7TeV.109308.AlpgenJimmyZmumubbNp3_nofilter				
Specifications	# events	$\sigma \times$ FE [pb]	k-factor	PDF	Transf. package
	9994	0.39	1.25	CTEQ6L1	16.6.5.5.1
Comments	MLM matching for partons with $p_T > 20$ GeV				

Table 45: $Z \to \mu^+\mu^- + b\bar{b} +$ jets (0...3 additional partons).

Process	$Z \to \tau^+\tau^- + b\bar{b} + 0p$				
Generator	ALPGEN + HERWIG/JIMMY				
DQ2 sample name	mc10_7TeV.109310.AlpgenJimmyZtautaubbNp0_nofilter				
Specifications	# events	$\sigma \times$ FE [pb]	k-factor	PDF	Transf. package
	149967	6.57	1.25	CTEQ6L1	16.6.5.5.1
Comments	MLM matching for partons with $p_T > 20\,\mathrm{GeV}$				

Process	$Z \to \tau^+\tau^- + b\bar{b} + 1p$				
Generator	ALPGEN + HERWIG/JIMMY				
DQ2 sample name	mc10_7TeV.109311.AlpgenJimmyZtautaubbNp1_nofilter				
Specifications	# events	$\sigma \times$ FE [pb]	k-factor	PDF	Transf. package
	99971	2.49	1.25	CTEQ6L1	16.6.5.5.1
Comments	MLM matching for partons with $p_T > 20\,\mathrm{GeV}$				

Process	$Z \to \tau^+\tau^- + b\bar{b} + 2p$				
Generator	ALPGEN + HERWIG/JIMMY				
DQ2 sample name	mc10_7TeV.109312.AlpgenJimmyZtautaubbNp2_nofilter				
Specifications	# events	$\sigma \times$ FE [pb]	k-factor	PDF	Transf. package
	39978	0.89	1.25	CTEQ6L1	16.6.5.5.1
Comments	MLM matching for partons with $p_T > 20\,\mathrm{GeV}$				

Process	$Z \to \tau^+\tau^- + b\bar{b} + 3p$				
Generator	ALPGEN + HERWIG/JIMMY				
DQ2 sample name	mc10_7TeV.109313.AlpgenJimmyZtautaubbNp3_nofilter				
Specifications	# events	$\sigma \times$ FE [pb]	k-factor	PDF	Transf. package
	8994	0.39	1.25	CTEQ6L1	16.6.5.5.1
Comments	MLM matching for partons with $p_T > 20\,\mathrm{GeV}$				

Table 46: $Z \to \tau^+\tau^- + b\bar{b}$ + jets (0...3 additional partons).

B LIST OF MONTE CARLO SAMPLES

Process	WW				
Generator	HERWIG				
DQ2 sample name	mc10_7TeV.105985.WW_Herwig				
Specifications	# events	$\sigma \times$ FE [pb]	k-factor	PDF	Transf. package
	249915	11.5003	1.48	MRST2007lomod	16.6.5.5.1
Comments	$W \to \ell\nu$, $(\ell = e, \mu, \tau)$ decay at least on lepton with $p_T(\ell) > 10\,\text{GeV}$ and $|\eta| < 2.8$				
Process	WZ				
Generator	HERWIG				
DQ2 sample name	mc10_7TeV.105987.WZ_Herwig				
Specifications	# events	$\sigma \times$ FE [pb]	k-factor	PDF	Transf. package
	249923	0.9722	1.30	MRST2007lomod	16.6.5.5.1
Comments	$W \to \ell\nu$ and $Z \to \ell^+\ell^-$ $(\ell = e, \mu, \tau)$ decays at least on lepton with $p_T(\ell) > 10\,\text{GeV}$ and $|\eta| < 2.8$				
Process	ZZ				
Generator	HERWIG				
DQ2 sample name	mc10_7TeV.105986.ZZ_Herwig				
Specifications	# events	$\sigma \times$ FE [pb]	k-factor	PDF	Transf. package
	249906	3.4641	1.60	MRST2007lomod	16.6.5.5.1
Comments	$Z \to \ell^+\ell^-$ $(\ell = e, \mu, \tau)$ decay at least on lepton with $p_T(\ell) > 10\,\text{GeV}$ and $|\eta| < 2.8$				

Table 47: Di-boson $(WW/WZ/ZZ)$ samples.

Process	$t(\bar{t}) + \bar{b}(b) \to be^+\nu_e(\bar{b}e^-\bar{\nu}_e) + \bar{b}(b)$ ($e\nu$ s-channel)				
Generator	MC@NLO + HERWIG/JIMMY				
DQ2 sample name	mc10_7TeV.108343.st_schan_enu_McAtNlo_Jimmy				
Specifications	# events	$\sigma \times$ FE [pb]	k-factor	PDF	Transf. package
	299831	0.47	1.00	CTEQ66	16.6.5.5.1
Comments	—				

Process	$t(\bar{t}) + \bar{b}(b) \to b\mu^+\nu_\mu(\bar{b}\mu^-\bar{\nu}_\mu) + \bar{b}(b)$ ($\mu\nu$ s-channel)				
Generator	MC@NLO + HERWIG/JIMMY				
DQ2 sample name	mc10_7TeV.108344.st_schan_munu_McAtNlo_Jimmy				
Specifications	# events	$\sigma \times$ FE [pb]	k-factor	PDF	Transf. package
	299877	0.47	1.00	CTEQ66	16.6.5.5.1
Comments	—				

Process	$t(\bar{t}) + \bar{b}(b) \to b\tau^+\nu_\tau(\bar{b}\tau^-\bar{\nu}_\tau) + \bar{b}(b)$ ($\tau\nu$ s-channel)				
Generator	MC@NLO + HERWIG/JIMMY				
DQ2 sample name	mc10_7TeV.108345.st_schan_taunu_McAtNlo_Jimmy				
Specifications	# events	$\sigma \times$ FE [pb]	k-factor	PDF	Transf. package
	299864	0.47	1.00	CTEQ66	16.6.5.5.1
Comments	—				

Process	$t(+b) \to e^+\nu_e b(+b)/\bar{t}(+\bar{b}) \to e^-\bar{\nu}_e \bar{b}(+\bar{b})$ ($e\nu$ t-channel)				
Generator	MC@NLO + HERWIG/JIMMY				
DQ2 sample name	mc10_7TeV.108340.st_tchan_enu_McAtNlo_Jimmy				
Specifications	# events	$\sigma \times$ FE [pb]	k-factor	PDF	Transf. package
	299897	7.12	1.00	CTEQ66	16.6.5.5.1
Comments	—				

Process	$t(+b) \to \mu^+\nu_\mu b(+b)/\bar{t}(+\bar{b}) \to \mu^-\bar{\nu}_\mu \bar{b}(+\bar{b})$ ($\mu\nu$ t-channel)				
Generator	MC@NLO + HERWIG/JIMMY				
DQ2 sample name	mc10_7TeV.108341.st_tchan_munu_McAtNlo_Jimmy				
Specifications	# events	$\sigma \times$ FE [pb]	k-factor	PDF	Transf. package
	299879	7.12	1.00	CTEQ66	16.6.5.5.1
Comments	—				

Process	$t(+b) \to \tau^+\nu_\tau b(+b)/\bar{t}(+\bar{b}) \to \tau^-\bar{\nu}_\tau \bar{b}(+\bar{b})$ ($\tau\nu$ t-channel)				
Generator	MC@NLO + HERWIG/JIMMY				
DQ2 sample name	mc10_7TeV.108342.st_tchan_taunu_McAtNlo_Jimmy				
Specifications	# events	$\sigma \times$ FE [pb]	k-factor	PDF	Transf. package
	299879	7.10	1.00	CTEQ66	16.6.5.5.1
Comments	—				

Process	$t(\bar{t}) + W^-(W^+)$ (Wt-channel)				
Generator	MC@NLO + HERWIG/JIMMY				
DQ2 sample name	mc10_7TeV.108346.st_Wt_McAtNlo_Jimmy				
Specifications	# events	$\sigma \times$ FE [pb]	k-factor	PDF	Transf. package
	899336	14.59	1.00	CTEQ66	16.6.5.5.1
Comments	—				

Table 48: Single top quark samples.

C Additional Plots

This part of the appendix contains additional plots that are of minor importance for the analysis presented in the thesis and are added for the sake of completeness.

Fig. 80 to Fig. 93 contain additional control plots for comparing data with MC expectation as presented in Sec. 10.8; Fig. 80 to Fig. 85 show distributions for the pretag samples, Fig. 86 to Fig. 93 contain the relevant plots after b-tagging respectively.

The fits to the invariant mass spectra of $Z \to ee$ and $Z \to e\gamma_{\text{fake}}$ in different bins of p_{T} and $|\eta|$ needed for the $e \to \gamma$ fake rate SFs evaluated in Sec. 11.4.1 are shown in Fig. 94 to Fig. 96.

In Sec. 11.4.2, the impact of the jet multiplicity on the fraction of prompt photons for the estimation of the W+jets+γ background had been studied and the result was shown in Fig. 63. The results of the template fits that enter Fig. 63 are depicted in Fig. 97 to Fig. 99.

Finally, Fig. 100 to Fig. 114 provide the complete set of the outcomes of the 3000 pseudo experiments for the estimation of the systematic uncertainties as described in Sec. 12.

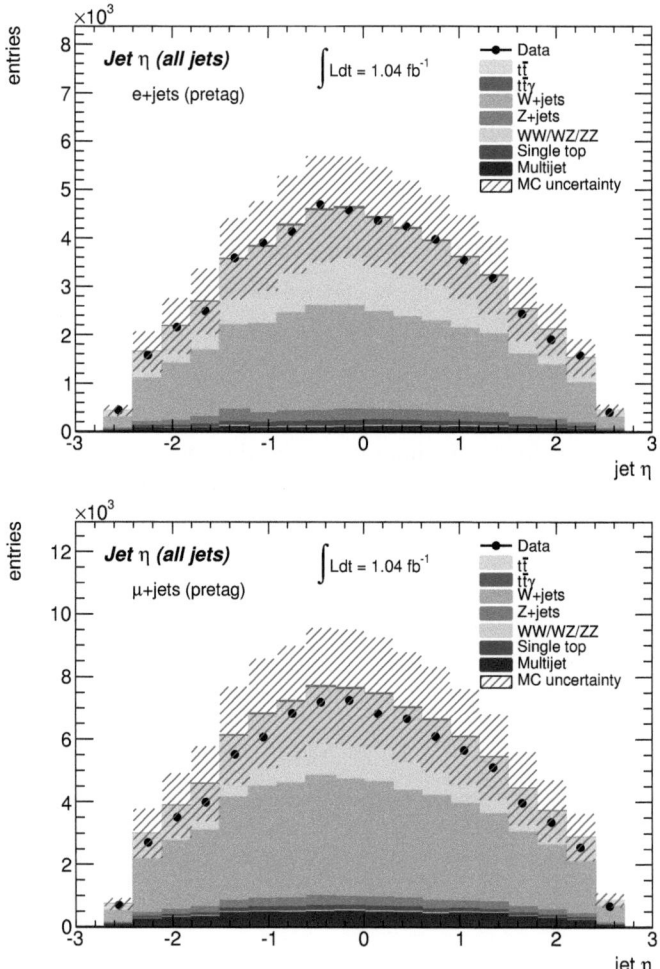

Figure 80: Jet η comparison between data and MC simulation for the event preselection before the b-tag requirement in the e+jets channel (upper plot) and the μ+jets channel (lower plot).

C ADDITIONAL PLOTS

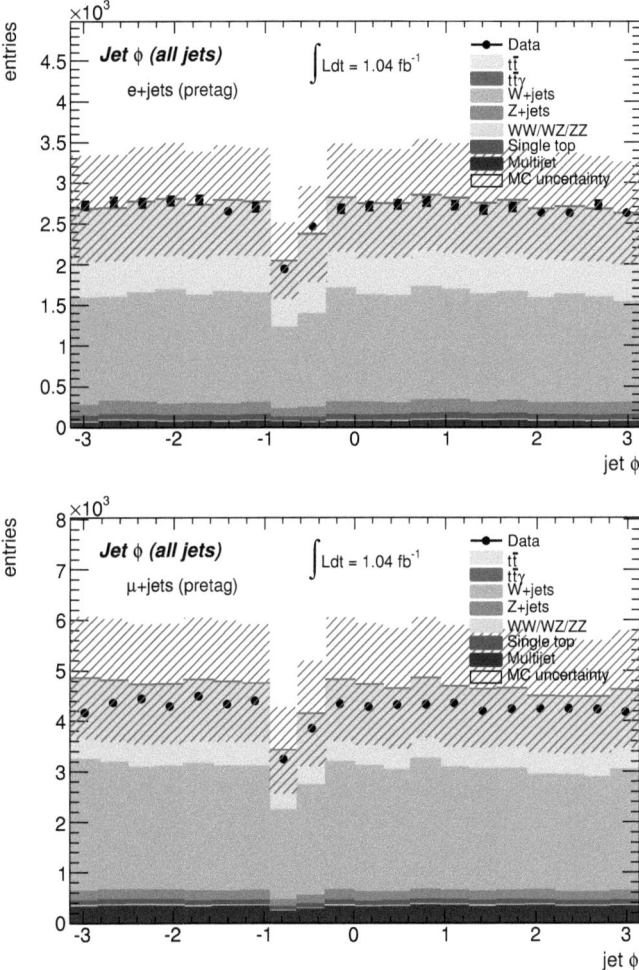

Figure 81: Jet ϕ comparison between data and MC simulation for the event preselection before the b-tag requirement in the e+jets channel (upper plot) and the μ+jets channel (lower plot).

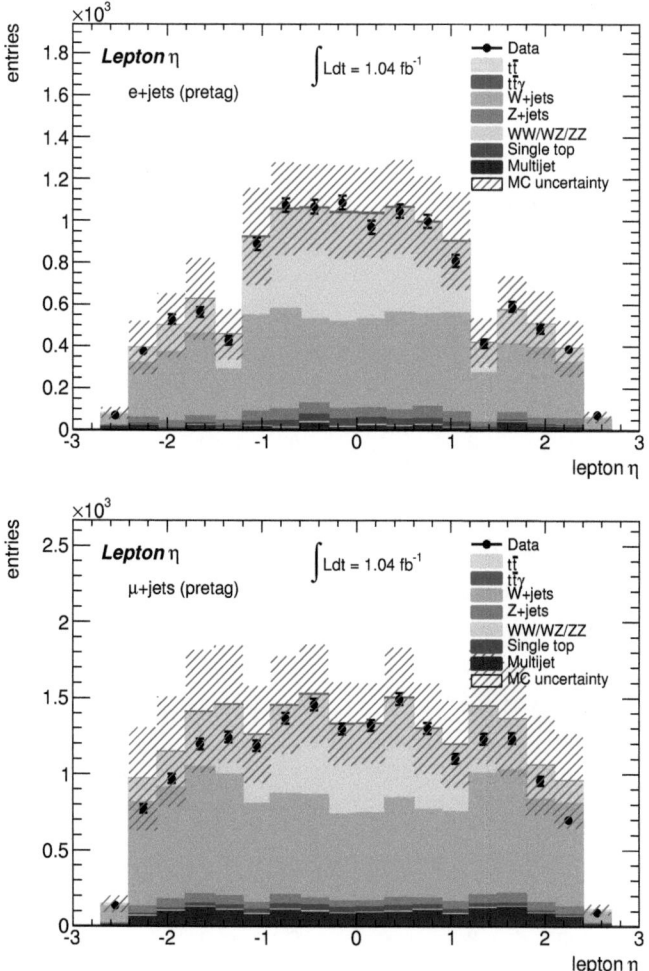

Figure 82: Lepton η comparison between data and MC simulation for the event preselection before the b-tag requirement in the e+jets channel (upper plot) and the μ+jets channel (lower plot).

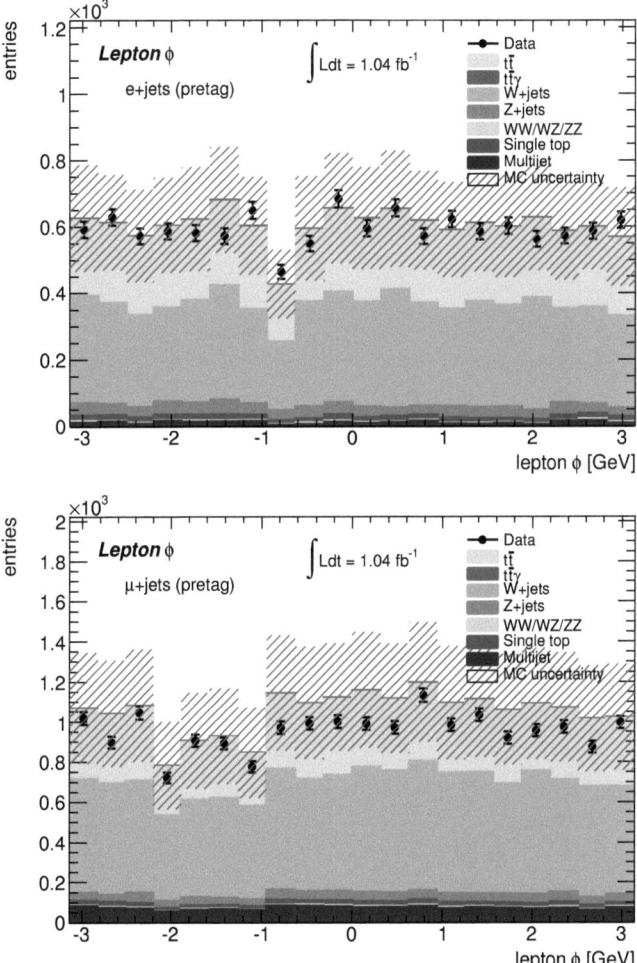

Figure 83: Lepton ϕ comparison between data and MC simulation for the event preselection before the b-tag requirement in the e+jets channel (upper plot) and the μ+jets channel (lower plot).

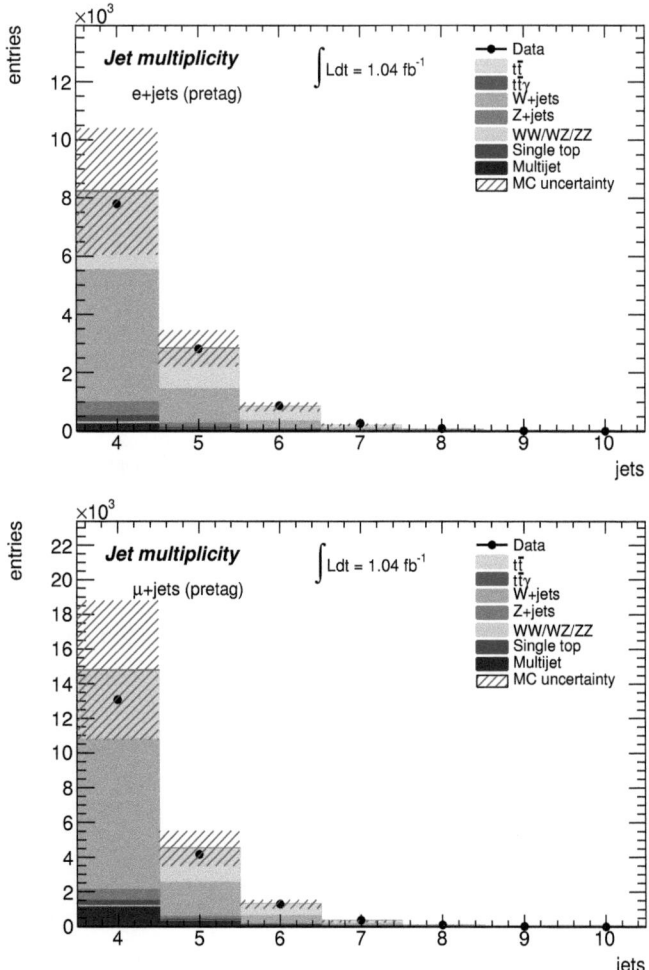

Figure 84: Jet multiplicity comparison between data and MC simulation for the event preselection before the b-tag requirement in the e+jets channel (upper plot) and the μ+jets channel (lower plot).

C ADDITIONAL PLOTS

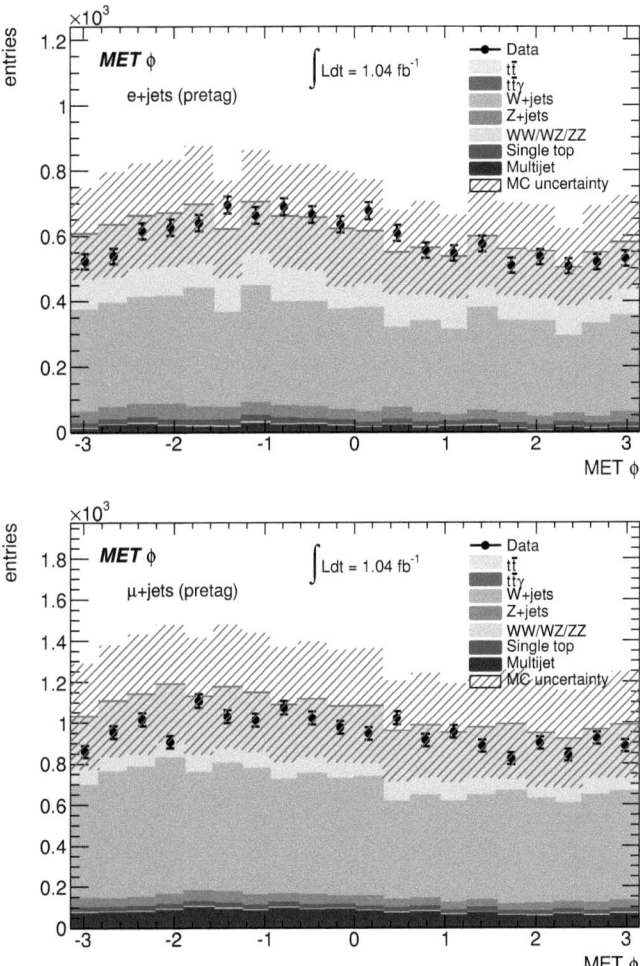

Figure 85: MET ϕ comparison between data and MC simulation for the event preselection before the b-tag requirement in the e+jets channel (upper plot) and the μ+jets channel (lower plot).

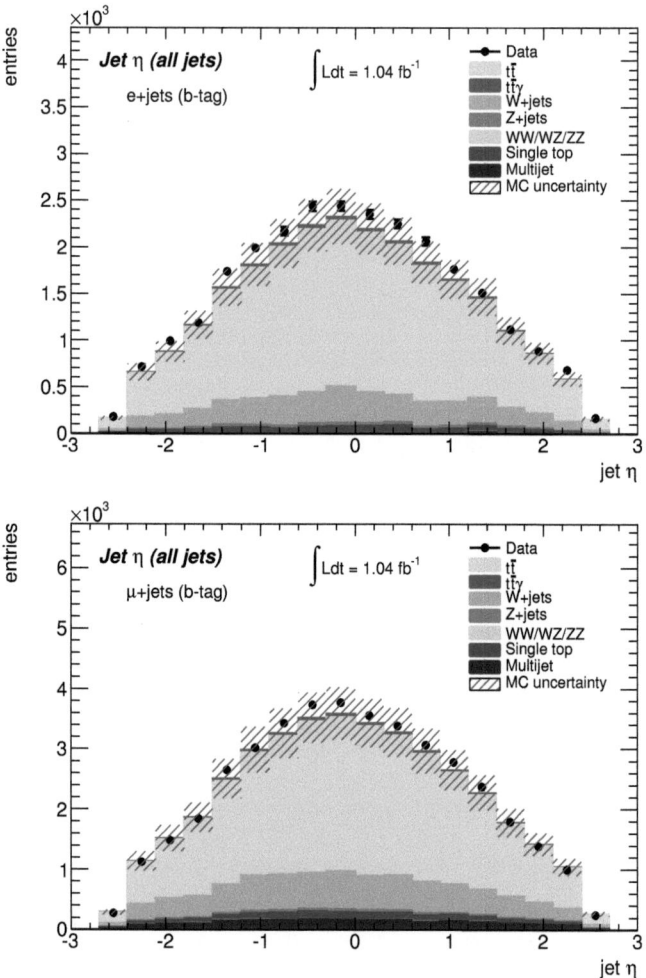

Figure 86: Jet η comparison between data and MC simulation for the event preselection after the b-tag requirement in the e+jets channel (upper plot) and the μ+jets channel (lower plot).

C ADDITIONAL PLOTS

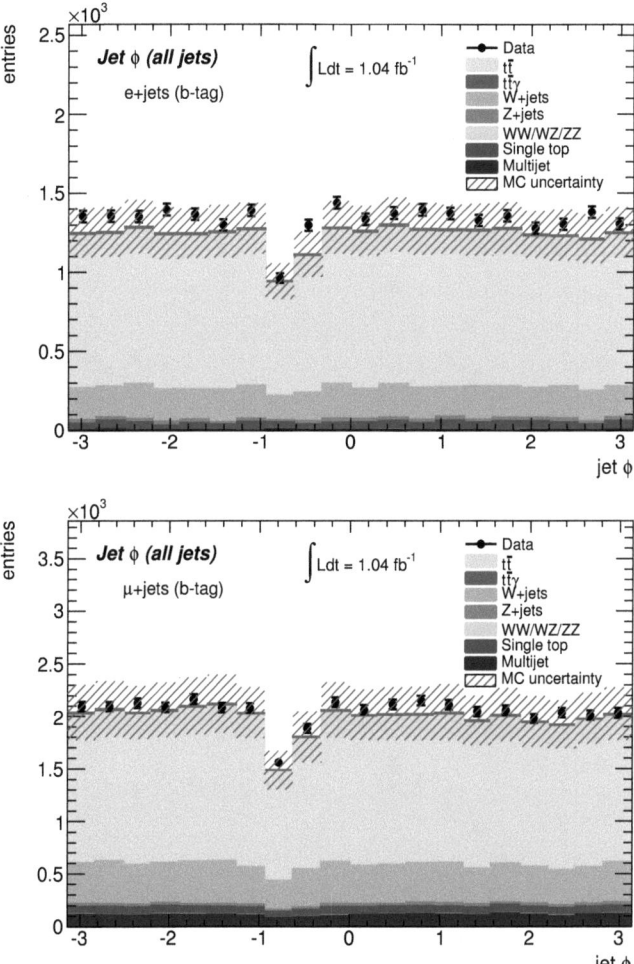

Figure 87: Jet ϕ comparison between data and MC simulation for the event preselection after the b-tag requirement in the e+jets channel (upper plot) and the μ+jets channel (lower plot).

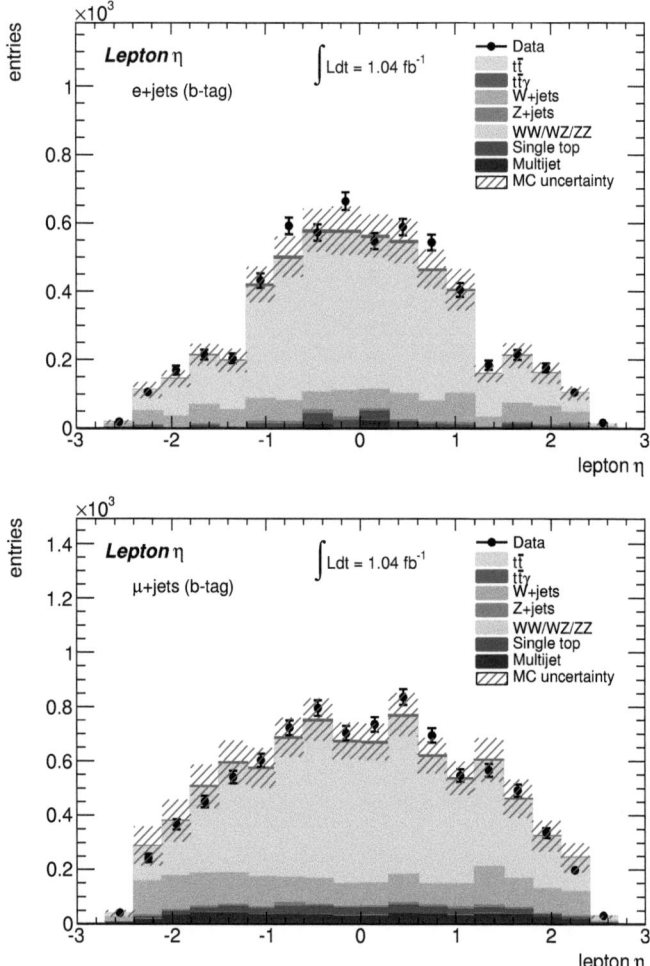

Figure 88: Lepton η comparison between data and MC simulation for the event preselection after the b-tag requirement in the e+jets channel (upper plot) and the μ+jets channel (lower plot).

C ADDITIONAL PLOTS

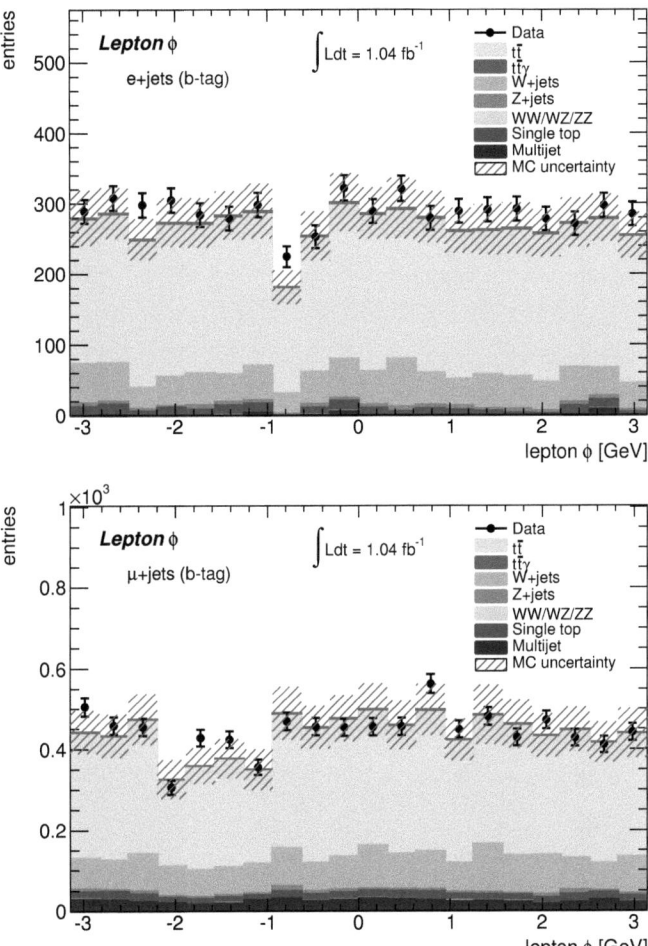

Figure 89: Lepton ϕ comparison between data and MC simulation for the event preselection after the b-tag requirement in the e+jets channel (upper plot) and the μ+jets channel (lower plot).

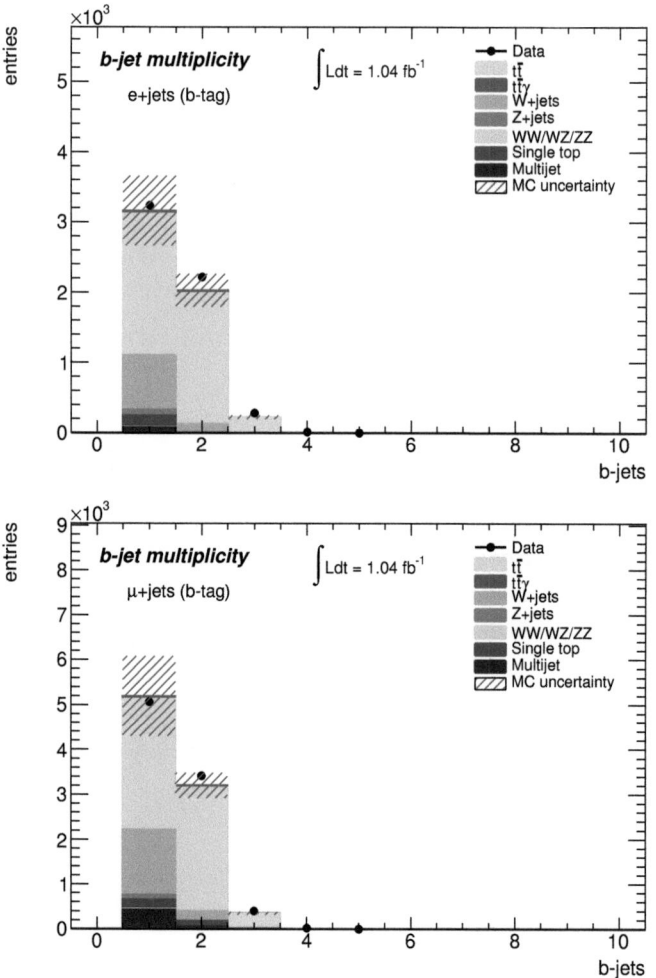

Figure 90: b-jet multiplicity comparison between data and MC simulation for the event preselection after the b-tag requirement in the e+jets channel (upper plot) and the μ+jets channel (lower plot).

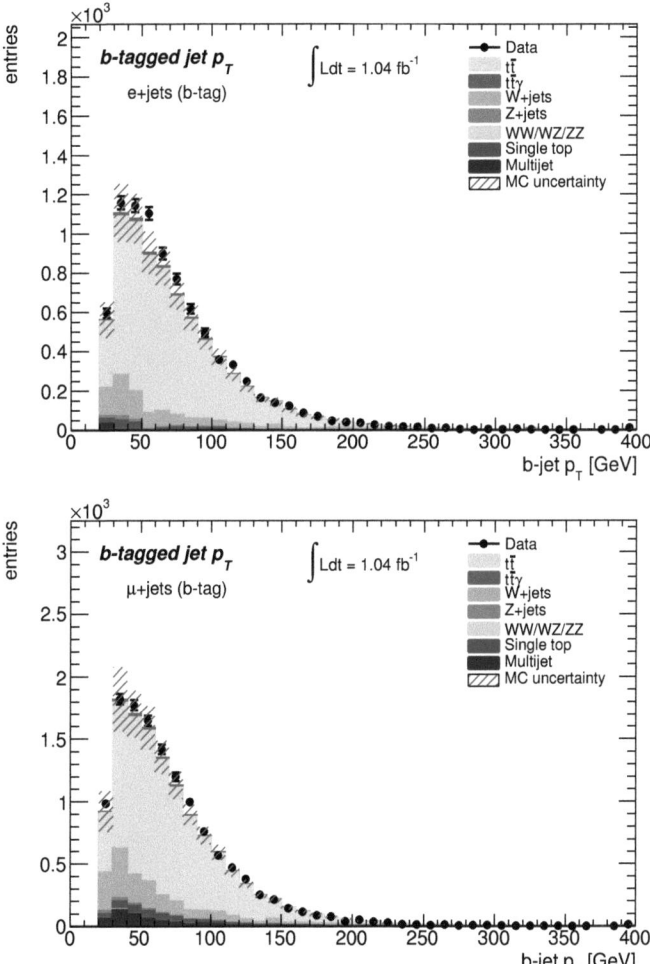

Figure 91: b-jet p_T comparison between data and MC simulation for the event preselection after the b-tag requirement in the e+jets channel (upper plot) and the μ+jets channel (lower plot).

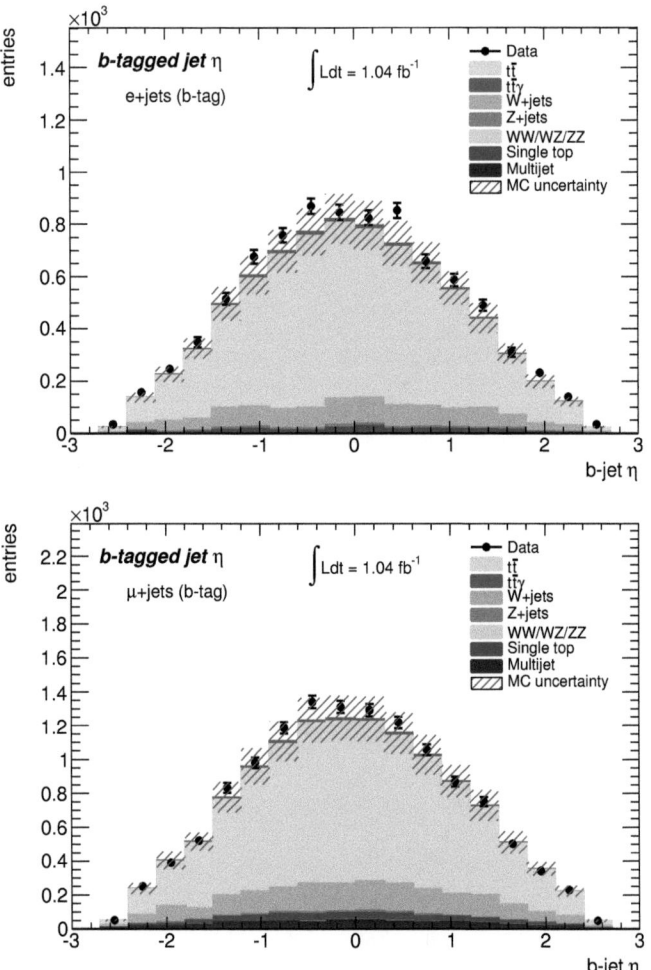

Figure 92: b-jet η comparison between data and MC simulation for the event preselection after the b-tag requirement in the e+jets channel (upper plot) and the μ+jets channel (lower plot).

C ADDITIONAL PLOTS

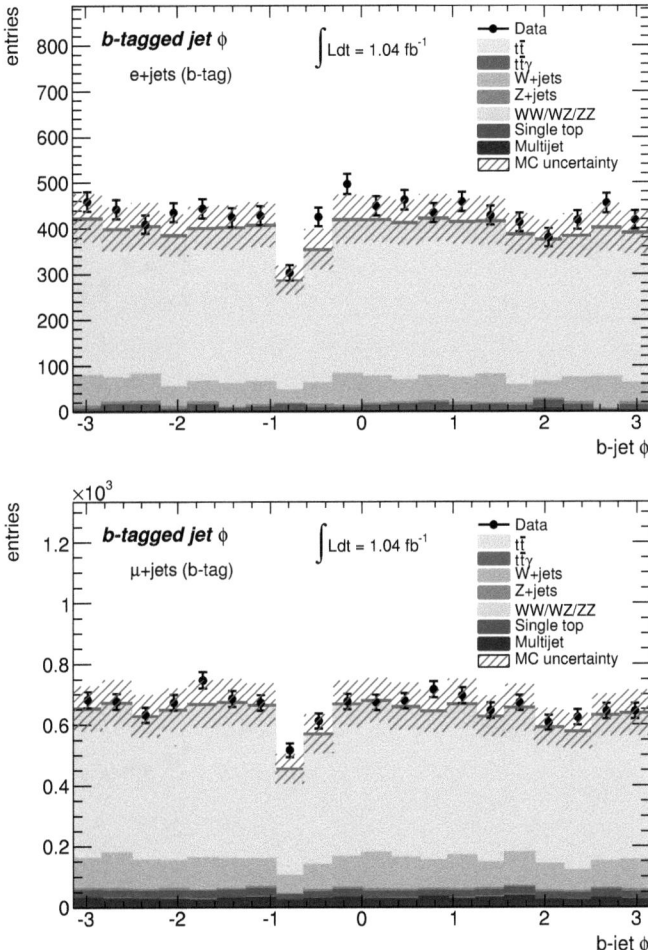

Figure 93: b-jet ϕ comparison between data and MC simulation for the event preselection after the b-tag requirement in the e+jets channel (upper plot) and the μ+jets channel (lower plot).

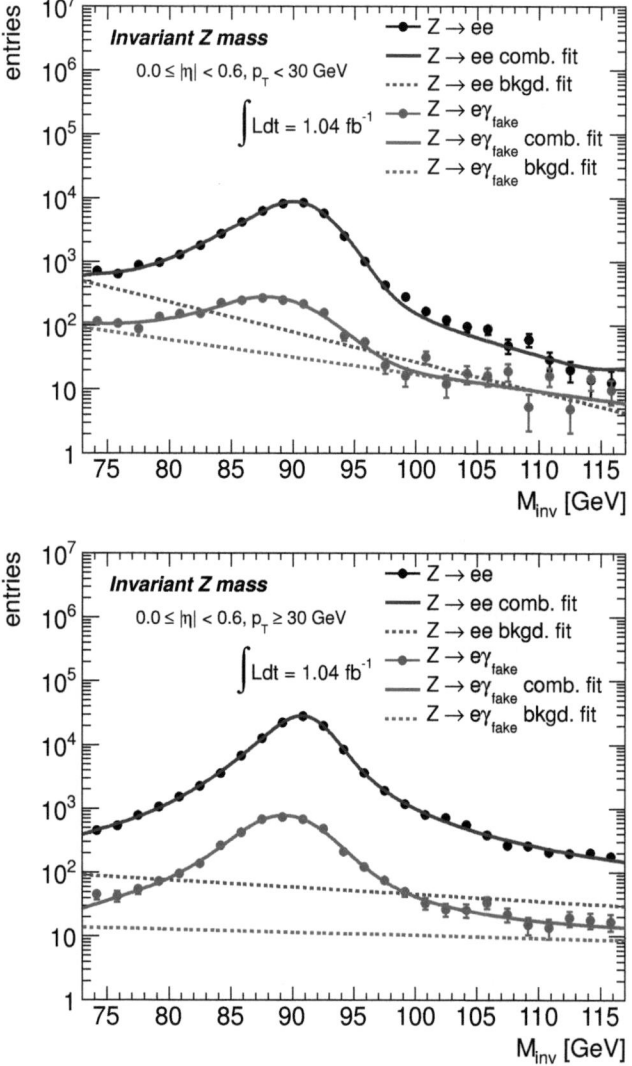

Figure 94: Invariant mass spectra $Z \to e^+e^-$ and $Z \to e\gamma_{\text{fake}}$ needed for the evaluation of the η-$_p$T dependent $e \to \gamma$ fake rate SFs in Sec. 11.4.1.

C ADDITIONAL PLOTS

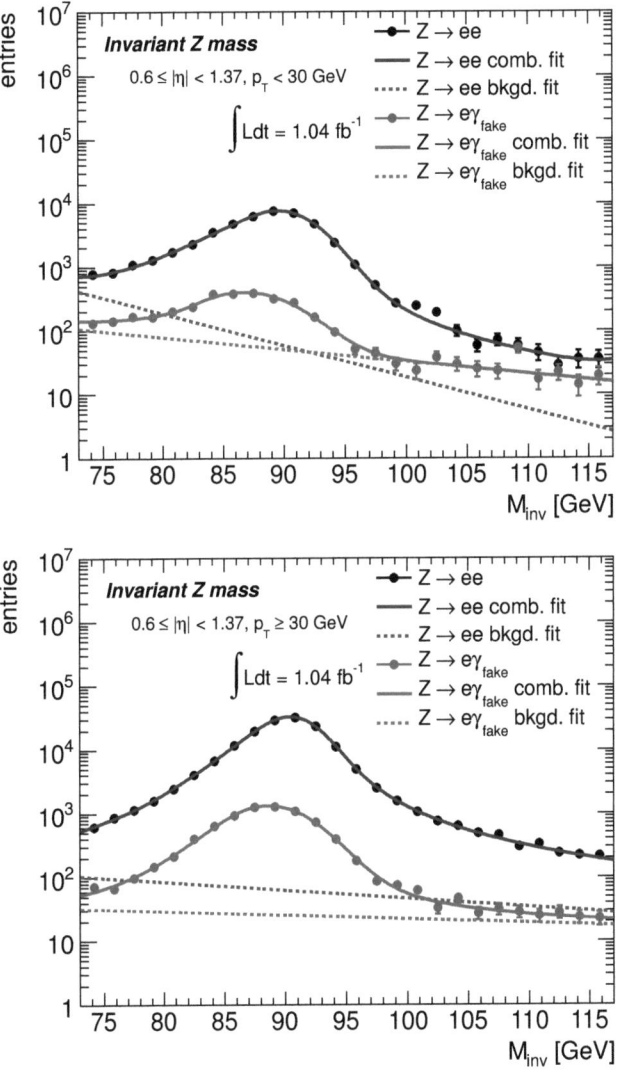

Figure 95: Invariant mass spectra $Z \to e^+e^-$ and $Z \to e\gamma_{\text{fake}}$ needed for the evaluation of the $\eta\text{-}p_T$ dependent $e \to \gamma$ fake rate SFs in Sec. 11.4.1.

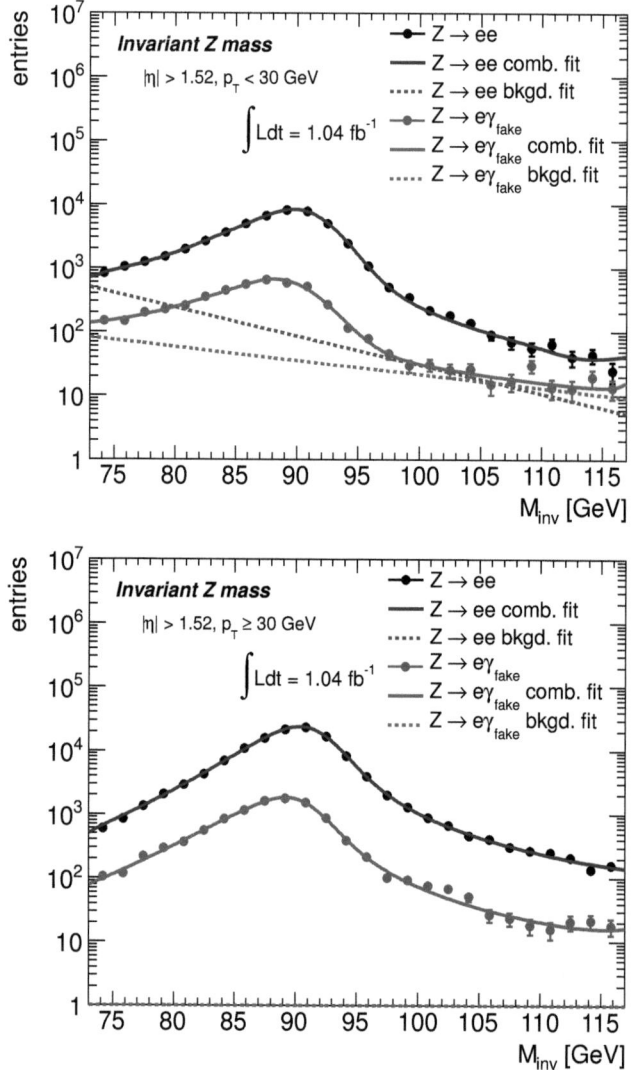

Figure 96: Invariant mass spectra $Z \to e^+e^-$ and $Z \to e\gamma_{\text{fake}}$ needed for the evaluation of the η-$_p$T dependent $e \to \gamma$ fake rate SFs in Sec. 11.4.1.

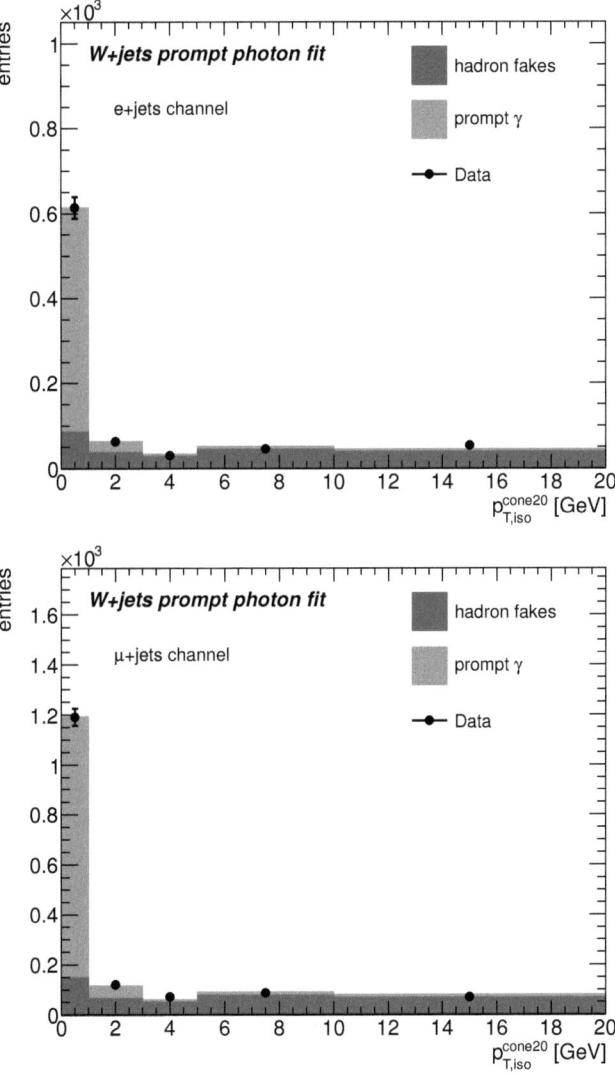

Figure 97: Template fit of $W + \text{jets} + \gamma$ in the CR for obtaining f_γ for different jet multiplicities.

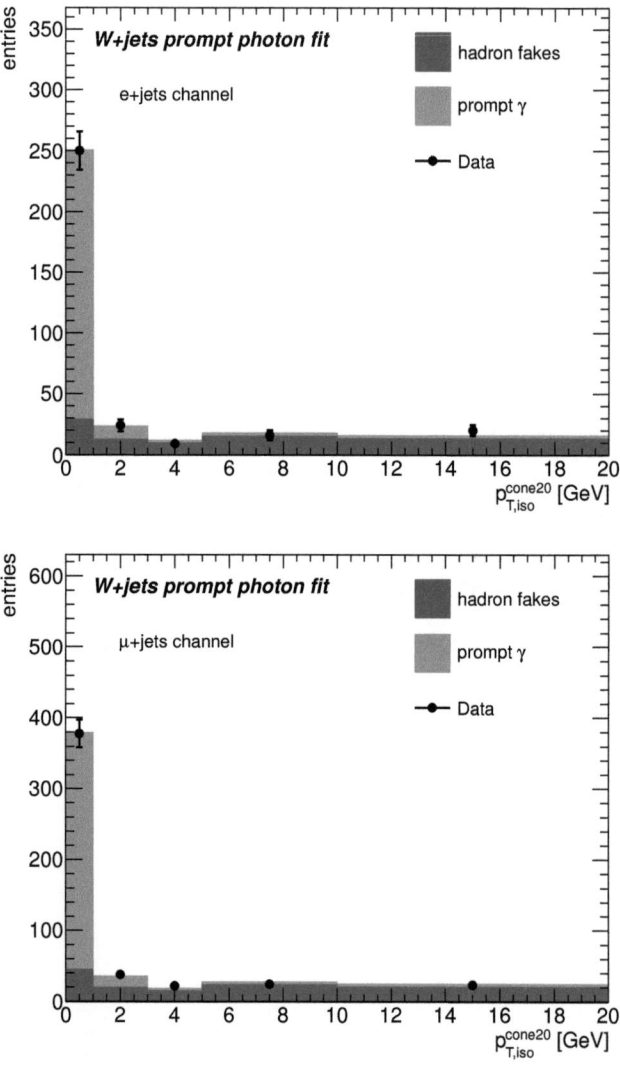

Figure 98: Template fit of $W + \text{jets} + \gamma$ in the CR for obtaining f_γ for different jet multiplicities.

C ADDITIONAL PLOTS

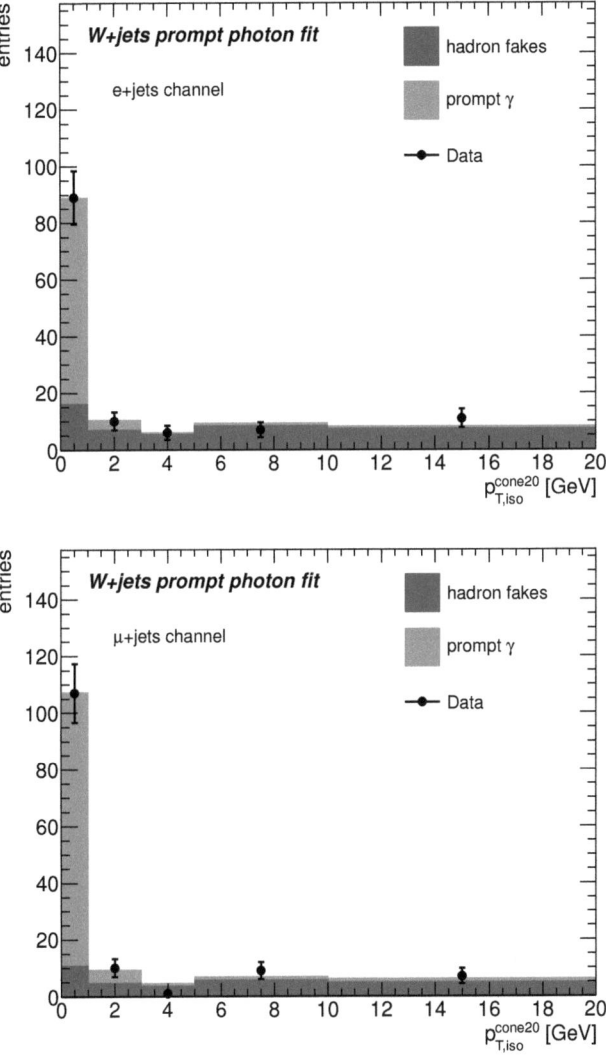

Figure 99: Template fit of $W+\text{jets}+\gamma$ in the CR for obtaining f_γ for different jet multiplicities.

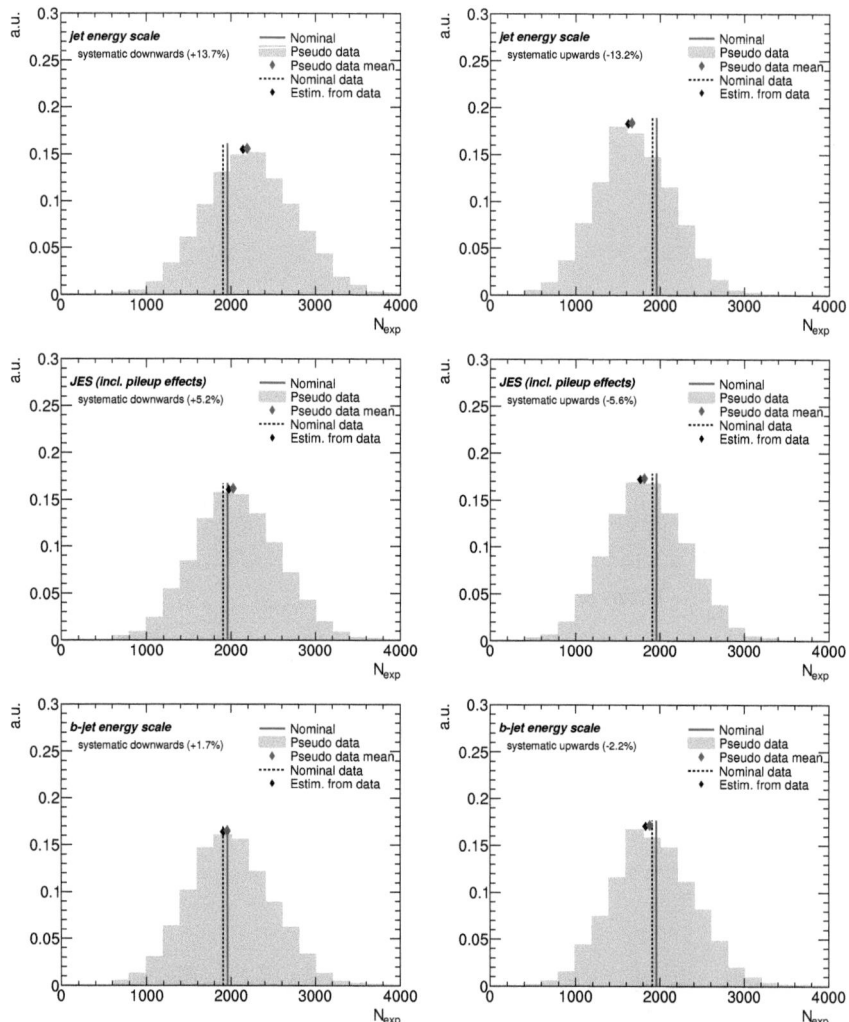

Figure 100: Systematic uncertainties related to jets (1).

C ADDITIONAL PLOTS

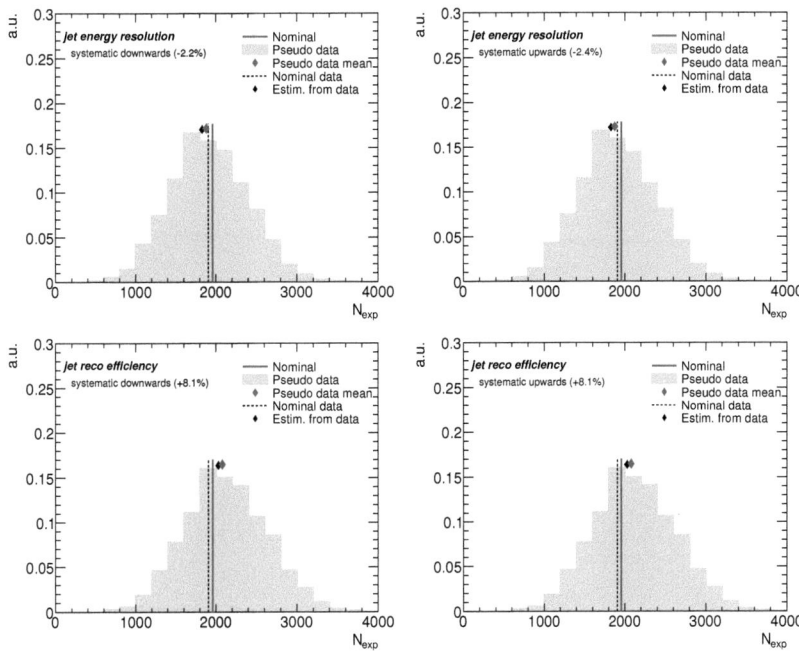

Figure 101: Systematic uncertainties related to jets (2).

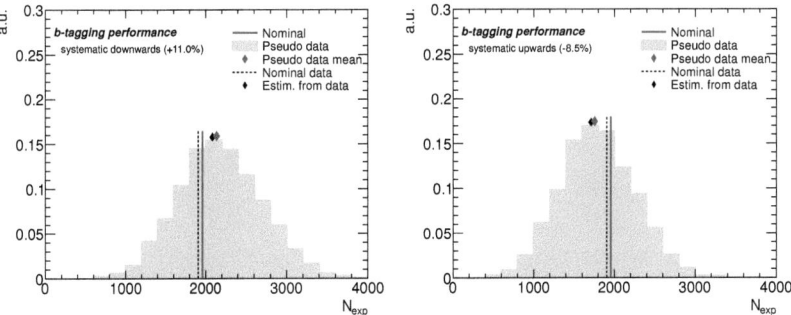

Figure 102: Systematic uncertainties related to b-tagging performance.

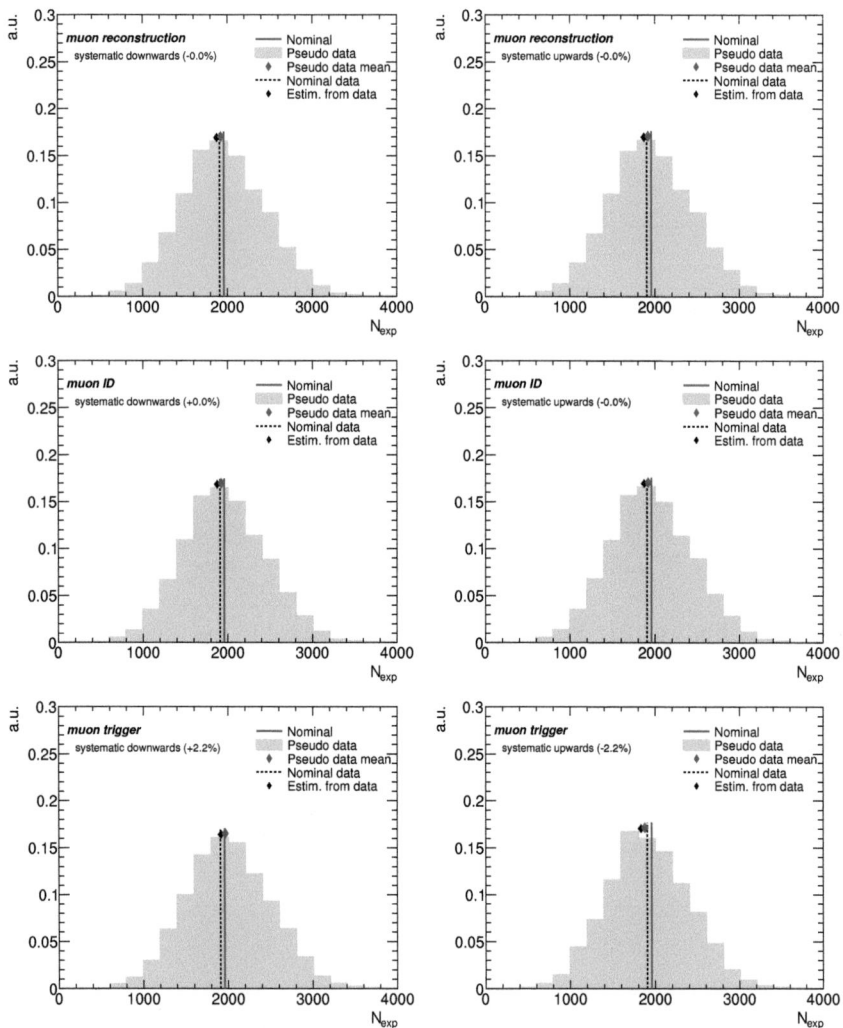

Figure 103: Systematic uncertainties related to muon performance (1).

C ADDITIONAL PLOTS

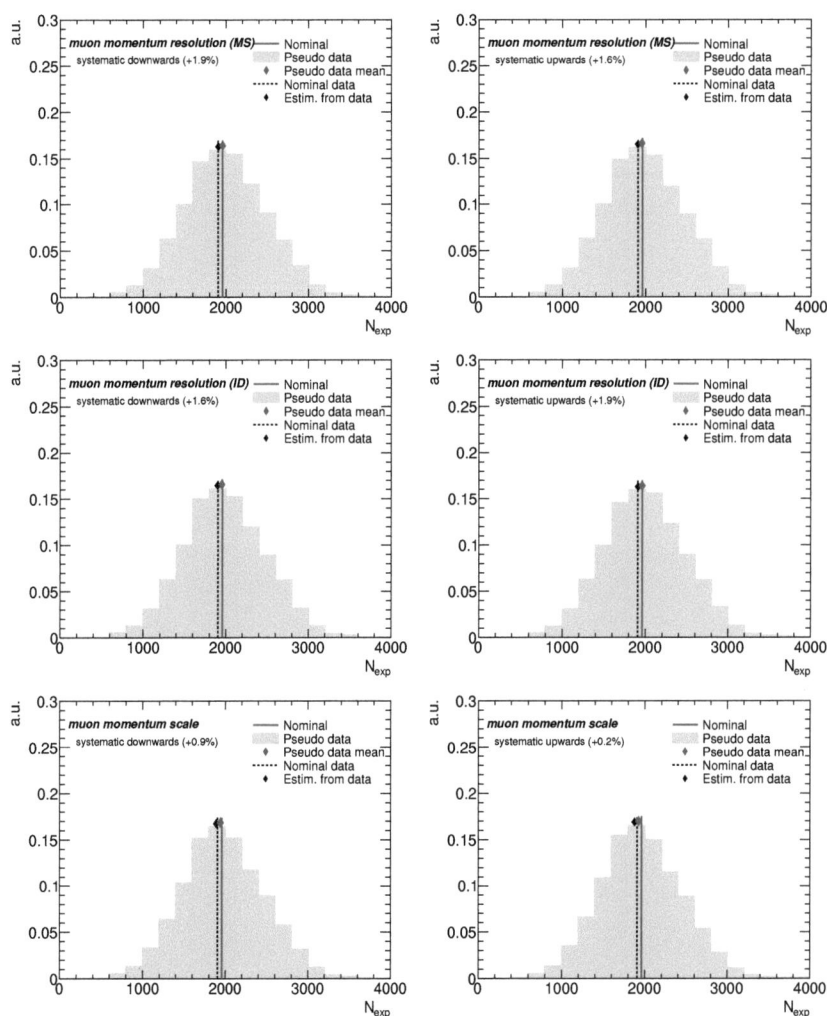

Figure 104: Systematic uncertainties related to muon performance (2).

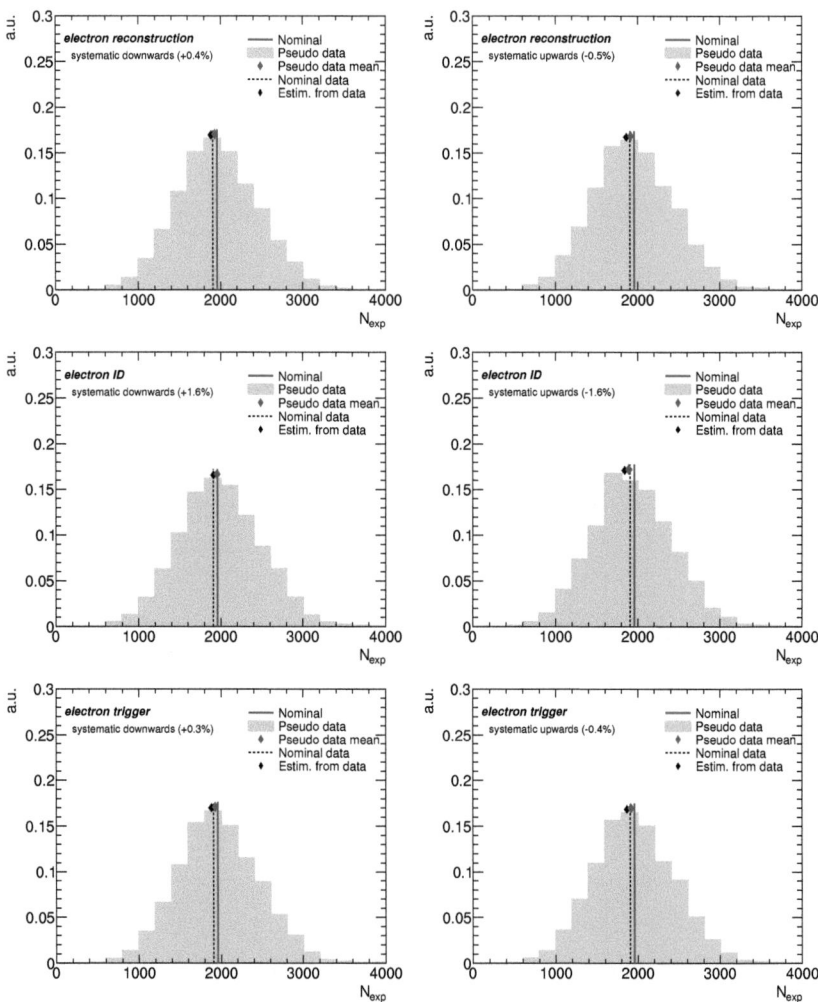

Figure 105: Systematic uncertainties related to electron performance (1).

C ADDITIONAL PLOTS

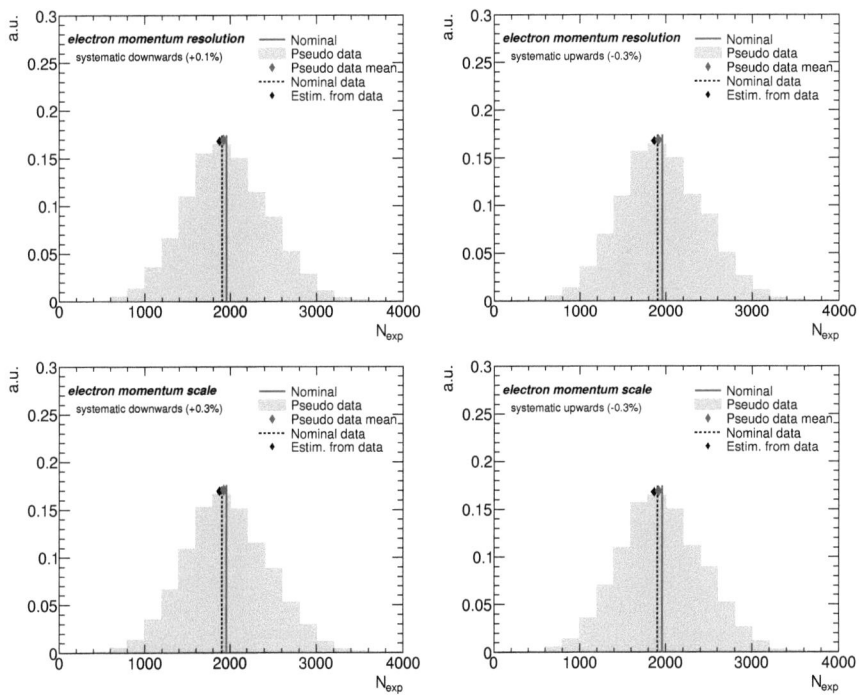

Figure 106: Systematic uncertainties related to electron performance (2).

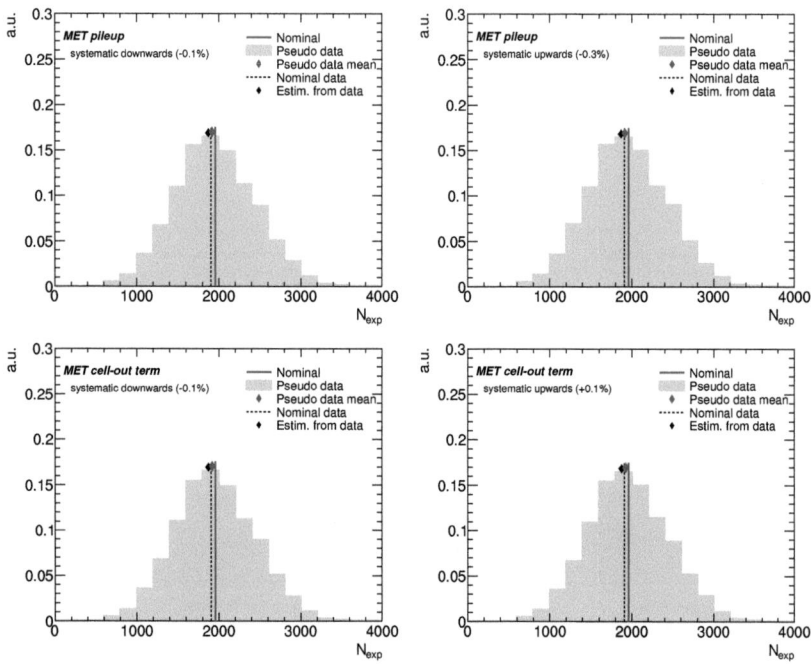

Figure 107: Systematic uncertainties related to $\not{E}_T$.

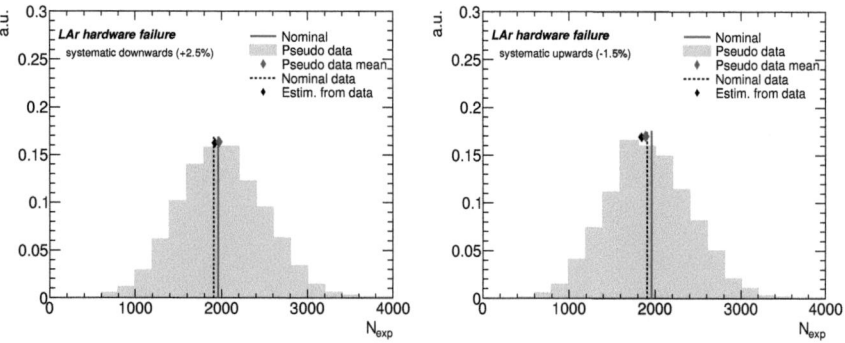

Figure 108: Systematic uncertainties related to the EMC hardware failure.

C ADDITIONAL PLOTS

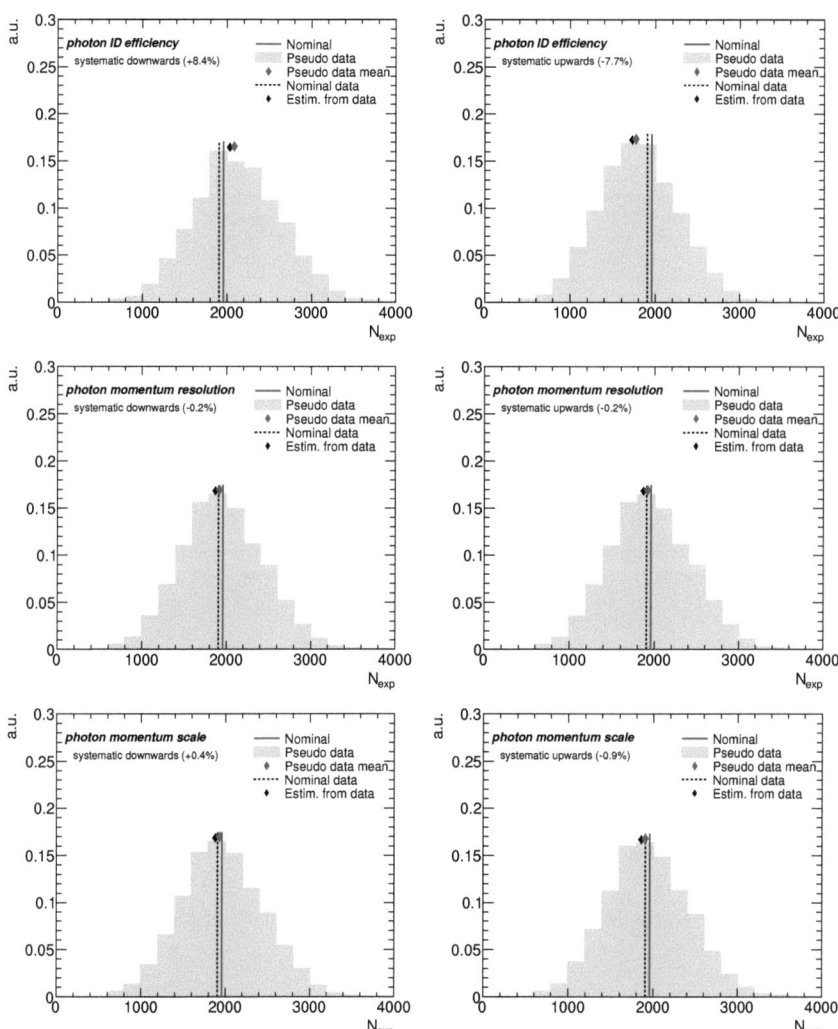

Figure 109: Systematic uncertainties related to photon performance.

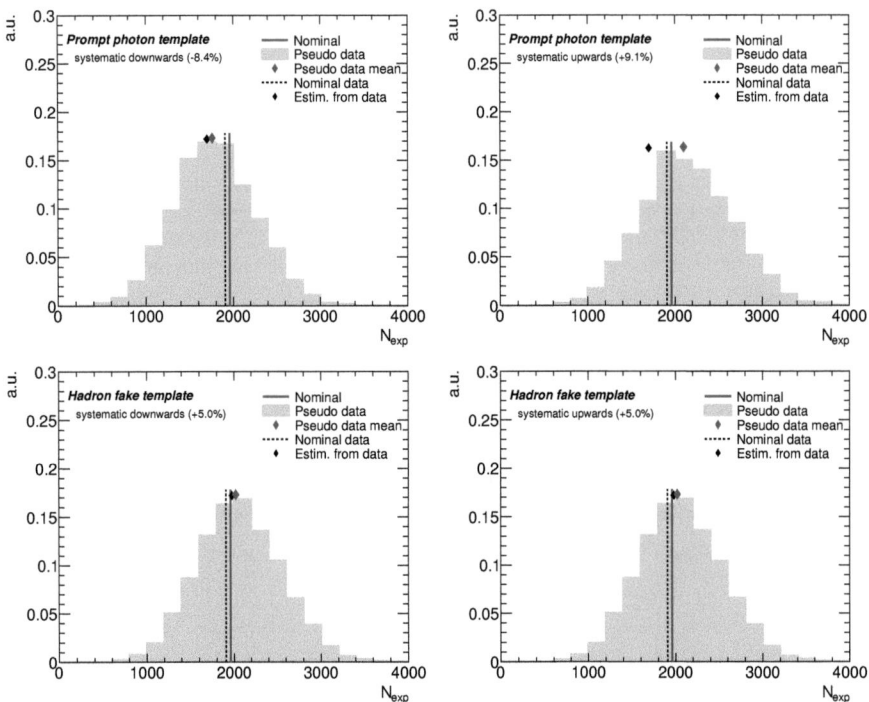

Figure 110: Systematic uncertainties related to the template modeling.

C ADDITIONAL PLOTS

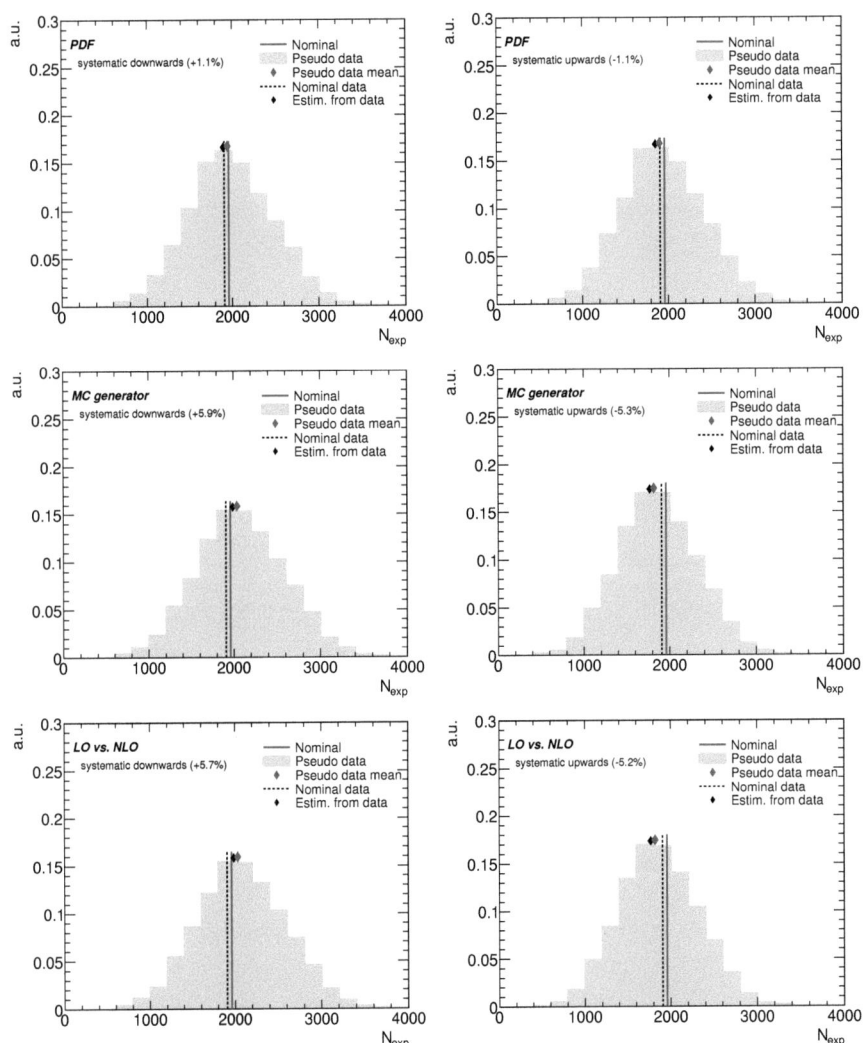

Figure 111: Systematic uncertainties related to MC simulation (1).

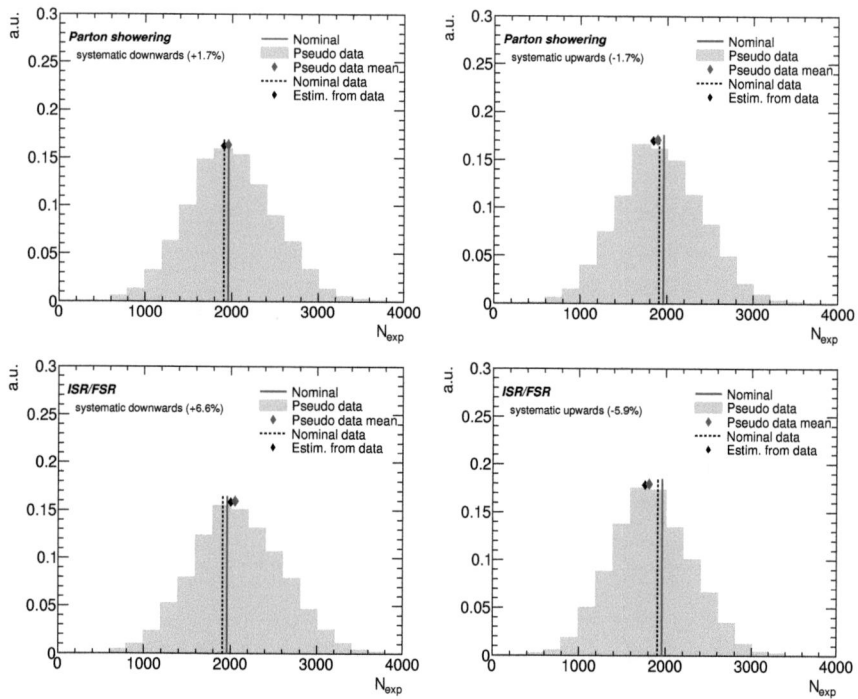

Figure 112: Systematic uncertainties related to MC simulation (2).

C ADDITIONAL PLOTS

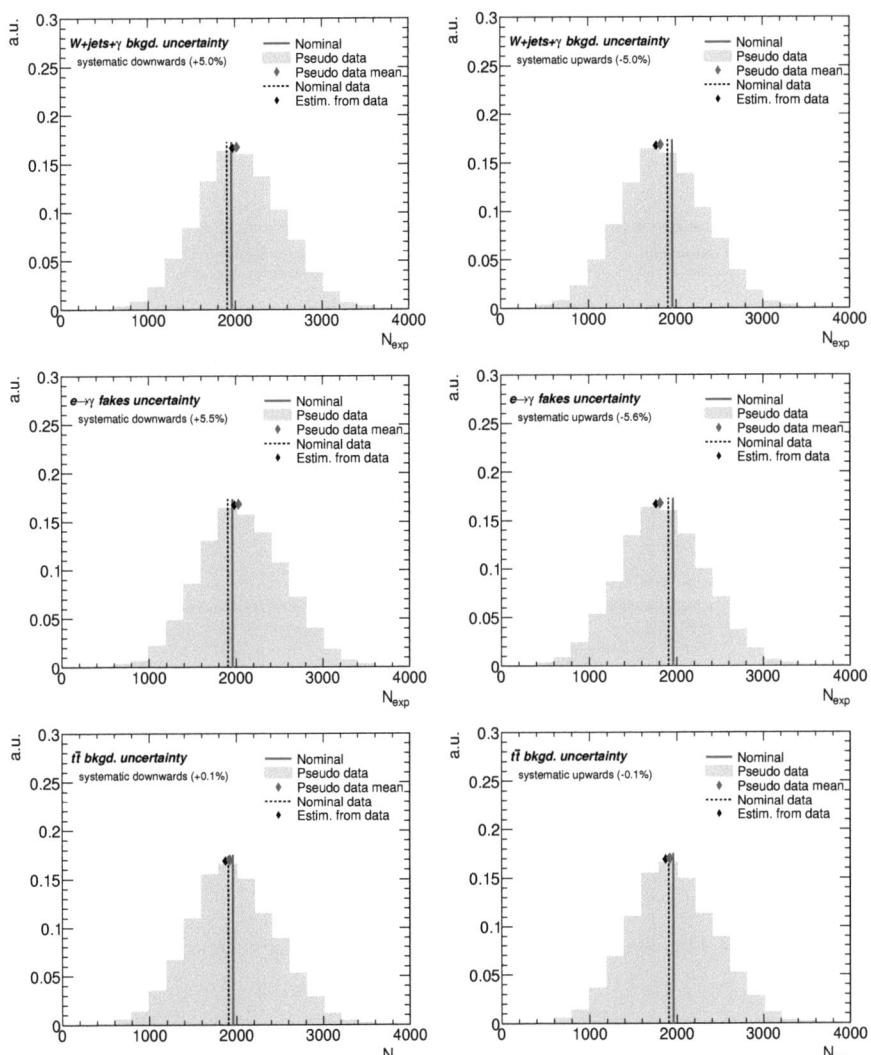

Figure 113: Systematic uncertainties related to background estimation (1).

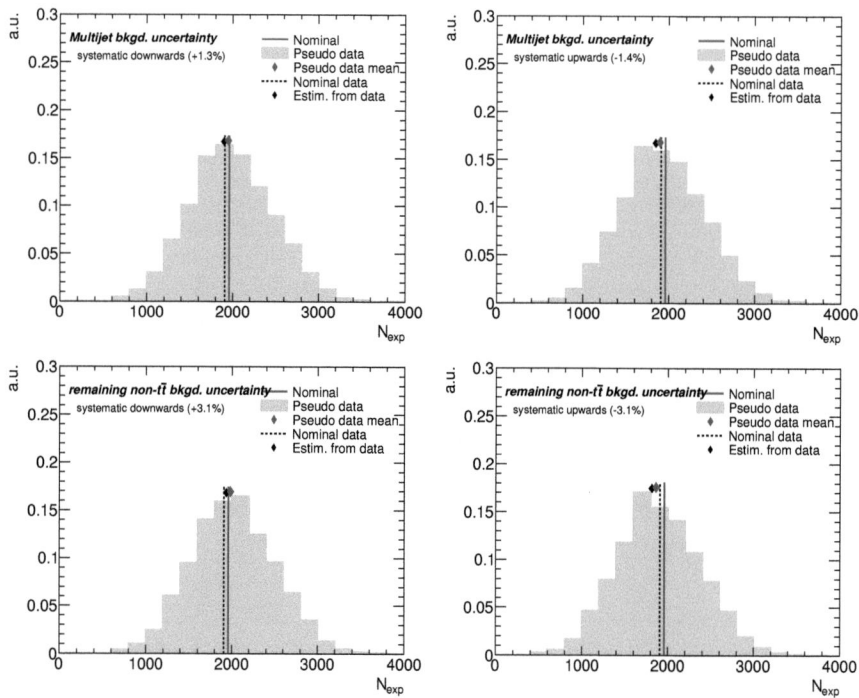

Figure 114: Systematic uncertainties related to background estimation (2).

References

[1] Michio Kaku. *Quantum Field Theory: A Modern Introduction.* Oxford University Press, 1993.

[2] P.A.M. Dirac. The Quantum Theory of the Emission and Absorption of Radiation. *Proc. Roy. Soc.*, **A114**, 243, 1927.

[3] J. Chadwick. The Intensity Distribution in the Magnetic Spectrum of β-Rays of Radium B + C. *Verhandl. Dtsch. Phys. Ges.*, **16**, 383, 1914.

[4] W. Pauli. Open Letter to Radioactive Persons. *Physics Today*, **31**, 27, 1930.

[5] W. Pauli. Septième Conseil de Physique. *Solvay (GauthierVillars, Paris)*, p. 324, 1934.

[6] E. Fermi. Towards the Theory of β-Rays. *Z. Phys.*, **88**, 161, 1934.

[7] G. Gamow and E. Teller. *Phys. Rev.*, **49**, 895, 1936.

[8] C. S. Wu et al. *Phys. Rev.*, **105**, 1413, 1957.

[9] J. Schwinger. *Ann. Phys.*, **2**, 407, 1957.

[10] T. D. Lee and C. N. Yang. *Phys. Rev.*, **108**, 1611, 1957.

[11] S.F. Novaes. Standard Model: An Introduction. 1999, arXiv:hep-ph/0001283.

[12] Paul Langacker. Introduction to the Standard Model and Electroweak Physics. pages 3–48, 2009, arXiv:0901.0241.

[13] The CMS Collaboration. Observation of a New Boson with a Mass near 125 GeV. 2012, CMS-PAS-HIG-12-020.

[14] Observation of an Excess of Events in the Search for the Standard Model Higgs Boson with the ATLAS Detector at the LHC. Technical Report ATLAS-CONF-2012-093, CERN, Geneva, Jul 2012.

[15] K. Nakamura et al. (Particle Data Group). *J. Phys. G* **37**, 075021, 2010.

[16] M. and Gell-Mann. A Schematic Model of Baryons and Mesons. *Physics Letters*, 8(3):214 – 215, 1964.

[17] Michael E. Peskin and Dan V. Schroeder. *An Introduction to Quantum Field Theory (Frontiers in Physics)*. Westview Press, 1995.

[18] N. Cabibbo. *Phys. Rev. Lett.* **10**, 531, 1963.

[19] M. Kobayashi and T. Maskawa. *Prog. Theor. Phys.*, **49**, 652, 1973.

[20] CKMfitter Group (J. Charles et al.). Predictions of Selected Flavour Observables within the Standard Model. *Eur. Phys. J. C41*, pages 1–131, 2011, arXiv:hep-ph/0406184.

[21] Ian J.R. Aitchison. Supersymmetry and the MSSM: An Elementary Introduction. 2005, arXiv:hep-ph/0505105.

[22] M.J. Duff. Kaluza-Klein Theory in Perspective. 1994, arXiv:hep-th/9410046.

[23] K. Nakamura et al. (Particle Data Group). *J. Phys. G* **37**, *075021*, 2010.

[24] F. Abe et al. Observation of Top Quark Production in $\bar{p}p$ Collisions. *Phys.Rev.Lett.*, 74:2626–2631, 1995.

REFERENCES

[25] S. Abachi et al. Observation of the Top Quark. *Phys.Rev.Lett.*, 74:2632–2637, 1995.

[26] Joseph R. Incandela, Arnulf Quadt, Wolfgang Wagner, and Daniel Wicke. Status and Prospects of Top Quark Physics. *Prog.Part.Nucl.Phys.*, 63:239–292, 2009.

[27] LHC Combination: Top Mass. (CMS-PAS-TOP-12-001), 2012.

[28] M. Beneke et al. Top Quark Physics. 2000, arXiv:hep-ph/0003033.

[29] Darwin Chang, We-Fu Chang, and Ernest Ma. Alternative Interpretation of the Tevatron Top Events. *Phys.Rev.*, D59:091503, 1999, arXiv:hep-ph/9810531.

[30] V.M. Abazov et al. Experimental Discrimination Between Charge $2e/3$ Top Quark and Charge $4e/3$ Exotic Quark Production Scenarios. *Phys.Rev.Lett.*, 98:041801, 2007, arXiv:hep-ex/0608044.

[31] Measurement of the Top Quark Charge in pp Collisions at $\sqrt{s} = 7\,\text{TeV}$ in the ATLAS Experiment. Technical Report ATLAS-CONF-2011-141, CERN, Geneva, Sep 2011.

[32] Stefano Forte. Parton Distributions at the Dawn of the LHC. *Acta Phys.Polon.*, B41:2859–2920, 2010, arXiv:1011.5247.

[33] A.D. Martin, W.J. Stirling, R.S. Thorne, and G. Watt. Parton Distributions for the LHC. *Eur.Phys.J.*, C63:189–285, 2009, arXiv:0901.0002.

[34] Raymond Brock et al. Handbook of Perturbative QCD. *Rev. Mod. Phys.*, 67:157, 1995.

[35] Celine Degrande, Jean-Marc Gerard, Christophe Grojean, Fabio Maltoni, and Geraldine Servant. Non-Resonant New Physics in Top Pair Production at Hadron Colliders. *JHEP*, 1103:125, 2011, arXiv:1010.6304.

[36] Measurement of the Top Quark Pair Production Cross-Section Based on a Statistical Combination of Measurements of Dilepton and Single-Lepton Final States at $\sqrt{s} = 7\,\text{TeV}$ with the ATLAS Detector. Technical Report ATLAS-CONF-2011-108, CERN, Geneva, Aug 2011.

[37] U. Baur, A. Juste, L.H. Orr, and D. Rainwater. Probing Electroweak Top Quark Couplings at Hadron Colliders. *Phys.Rev.*, D71:054013, 2005, arXiv:hep-ph/0412021.

[38] J.A. Aguilar-Saavedra. A Minimal Set of Top Anomalous Couplings. *Nucl.Phys.*, B812:181–204, 2009, arXiv:0811.3842.

[39] T. Aaltonen et al. Evidence for $t\bar{t}\gamma$ Production and Measurement of $\sigma_{t\bar{t}\gamma}/\sigma_{t\bar{t}}$. *Phys.Rev.*, D84:031104, 2011, arXiv:1106.3970.

[40] Oliver Sim Brüning, Paul Collier, P Lebrun, Stephen Myers, Ranko Ostojic, John Poole, and Paul Proudlock. *LHC Design Report Vol. 2*. CERN, Geneva, 2004.

[41] Oliver Sim Brüning, Paul Collier, P Lebrun, Stephen Myers, Ranko Ostojic, John Poole, and Paul Proudlock. *LHC Design Report Vol. 1*. CERN, Geneva, 2004.

[42] Maximilien Brice. Views of the LHC Tunnel Sector 3-4. http://cdsweb.cern.ch/record/1211045, Oct 2009.

[43] Christiane Lefvre. The CERN Accelerator Complex. http://cdsweb.cern.ch/record/1260465, Dec 2008.

[44] AC Team. Diagram of an LHC Dipole Magnet. http://cdsweb.cern.ch/record/40524, Jun 1999.

[45] Michael Benedikt, Paul Collier, V Mertens, John Poole, and Karlheinz Schindl. *LHC Design Report Vol. 3*. CERN, Geneva, 2004.

[46] S van der Meer. Calibration of the Effective Beam Height in the ISR. Technical Report CERN-ISR-PO-68-31. ISR-PO-68-31, CERN, Geneva, 1968.

[47] Luminosity Determination in pp Collisions at $\sqrt{s} = 7\,\text{TeV}$ using the ATLAS Detector in 2011. Technical Report ATLAS-CONF-2011-116, CERN, Geneva, Aug 2011.

[48] Doug Schouten and Michel Vetterli. In-Situ Jet Calibration and the Effects of Pileup in ATLAS. Technical Report ATL-PHYS-INT-2007-011. ATL-COM-PHYS-2007-057, CERN, Geneva, Sep 2007.

[49] Joao Pequenao. CERN-GE-0803012 01. http://cdsweb.cern.ch/record/1095924, 2008.

[50] Joao Pequena. CERN-GE-0803022 01. http://cdsweb.cern.ch/record/1096081, 2008.

[51] *ATLAS Detector and Physics Performance: Technical Design Report, 1*. Technical Design Report ATLAS. CERN, Geneva, 1999.

[52] CERN. Geneva. LHC Experiments Committee; LHCC. ATLAS Inner Detector: Technical Design Report. Vol. 1. 1997. CERN-LHCC-97-16.

[53] Joao Pequena. CERN-GE-0803014 01. http://cdsweb.cern.ch/record/1095926, 2008.

[54] Norbert Wermes and G Hallewel. *ATLAS Pixel Detector: Technical Design Report*. Technical Design Report ATLAS. CERN, Geneva, 1998.

[55] Joao Pequena. CERN-GE-0803013 02.
http://cdsweb.cern.ch/record/1095925, 2008.

[56] Joao Pequena. CERN-GE-0803015 01.
http://cdsweb.cern.ch/record/1095927, 2008.

[57] CERN-EX-9308048_09 1.
http://cdsweb.cern.ch/record/39737, 1993.

[58] *ATLAS Liquid-Argon Calorimeter: Technical Design Report*. Technical Design Report ATLAS. CERN, Geneva, 1996.

[59] *ATLAS Central Solenoid: Technical Design Report*. Technical Design Report ATLAS. CERN, Geneva, 1997.

[60] Maximilien Brice. Installing the ATLAS Calorimeter: Vue Centrale du Détecteur ATLAS avec ses huit Toroides Entourant le Calorimètre avant son Déplacement au Centre du Détecteur, Nov 2005.

[61] Christina, Aurelien Muller, and Maximiliem Brice. Preparation for the Lowering of the Second Magnet End Cap Toroid into ATLAS Cavern Side C, Jul 2007.

[62] D.A. Scannicchio. ATLAS Trigger and Data Acquisition: Capabilities and Commissioning. *Nucl. Instr.*, 617, 1-3:306–309, May 2010.

[63] *ATLAS Computing: Technical Design Report*. CERN, Geneva, 2005.

[64] F. Rademakers and René Brun. ROOT: An Object-Oriented Data Analysis Framework. *Linux Journal*, 1998.

[65] S. Frixione and B.R. Webber. Matching NLO QCD Computations and Parton Shower Simulations. *JHEP*, 0206:029, 2002, arXiv:hep-ph/0204244.

[66] S. Frixione, P. Nason and C. Oleari. Matching NLO QCD Computations with Parton Shower Simulations: the POWHEG Method. *JHEP*, 0711:070, 2007, arXiv:0709.2092.

[67] M.L. Mangano et al. ALPGEN, a Generator for Hard Multiparton Processes in Hadronic Collisions. *JHEP*, 0307:001, 2003, arXiv:hep-ex/0206293.

[68] B.P. Kersevan and E. Richter-Wąs. The Monte Carlo Event Generator AcerMC Version 2.0 with Interfaces to PYTHIA 6.2 and HERWIG 6.5. 2004, arXiv:hep-ph/0405247.

[69] Wolfgang Kilian, Thorsten Ohl, and Jurgen Reuter. WHIZARD: Simulating Multi-Particle Processes at LHC and ILC. *Eur.Phys.J.*, C71:1742, 2011, arXiv:0708.4233.

[70] Mauro Moretti, Thorsten Ohl, and Jurgen Reuter. O'Mega: An Optimizing Matrix Element Generator. 2001, arXiv:hep-ph/0102195.

[71] Francesco Caravaglios and Mauro Moretti. An Algorithm to Compute Born Scattering Amplitudes without Feynman Graphs. *Phys.Lett.*, B358:332–338, 1995, arXiv:hep-ph/9507237.

[72] Thorsten Ohl. Vegas Revisited: Adaptive Monte Carlo Integration Beyond Factorization. *Comput.Phys.Commun.*, 120:13–19, 1999, arXiv:hep-ph/9806432.

[73] G. Peter Lepage. VEGAS: an Adaptive Multidimensional Integration Program. 1980.

[74] G. Corcella et al. HERWIG 6: An Event Generator for Hadron Emission Reactions with Interfering Gluons (Including Supersymmetric Processes). *JHEP*, 0101:010, 2001, arXiv:hep-ph/0011363.

[75] T. Sjostrand, S. Mrenna, and P.Z. Skands. PYTHIA 6.4 Physics and Manual. *JHEP*, 05:026, 2006, arXiv:hep-ph/0603175.

[76] Bo Andersson, G. Gustafson, G. Ingelman, and T. Sjostrand. Parton Fragmentation and String Dynamics. *Phys.Rept.*, 97:31–145, 1983.

[77] Piotr Golonka and Zbigniew Was. PHOTOS Monte Carlo: A Precision Tool for QED Corrections in Z and W Decays. *Eur.Phys.J.*, C45:97–107, 2006, arXiv:hep-ph/0506026.

[78] J. M. Butterworth, Jeffrey R. Forshaw, and M. H. Seymour. Multiparton Interactions in Photoproduction at HERA. *Z. Phys.*, C72:637–646, 1996, arXiv:hep-ph/9601371.

[79] Expected Electron Performance in the ATLAS Experiment. Technical Report ATL-PHYS-PUB-2011-006, CERN, Geneva, Apr 2011.

[80] Muon Reconstruction Efficiency in Reprocessed 2010 LHC Proton-Proton Collision Data Recorded with the ATLAS Detector. Technical Report ATLAS-CONF-2011-063, CERN, Geneva, Apr 2011.

[81] D Orestano and W Liebig. Muon Performance in Minimum Bias pp Collision Data at $\sqrt{s} = $ 7TeV with ATLAS. Technical Report ATLAS-COM-CONF-2010-036, CERN, Geneva, May 2010.

[82] L Asquith, B Brelier, J Butterworth, et al. Performance of Jet Algorithms in the ATLAS Detector. Technical Report ATL-COM-PHYS-2009-630, CERN, Geneva, Dec 2009.

[83] Matteo Cacciari, Gavin P. Salam, and Gregory Soyez. The anti-k_t Jet Clustering Algorithm. *JHEP*, page 063.

[84] Properties of Jets and Inputs to Jet Reconstruction and Calibration with the ATLAS Detector Using Proton-Proton Collisions at $\sqrt{s} = 7\,\text{TeV}$. Technical Report ATLAS-CONF-2010-053, CERN, Geneva, Jul 2010.

[85] Jet Energy Scale and its Systematic Uncertainty in Proton-Proton Collisions at $\sqrt{s} = 7\,\text{TeV}$ in ATLAS 2010 Data. Technical Report ATLAS-CONF-2011-032, CERN, Geneva, Mar 2011.

[86] J. Beringer et al. (Particle Data Group). *Phys. Rev. D86*, 010001, 2012.

[87] G. Aad et al. Expected Performance of the ATLAS Experiment - Detector, Trigger and Physics. 2009, 0901.0512.

[88] G Piacquadio and C Weiser. A New Inclusive Secondary Vertex Algorithm for b-Jet Tagging in ATLAS. *Journal of Physics: Conference Series*, 119(3):032032, 2008.

[89] Commissioning of the ATLAS High-Performance b-Tagging Algorithms in the 7 TeV Collision Data. Technical Report ATLAS-CONF-2011-102, CERN, Geneva, Jul 2011.

[90] B Alvarez, V Boisvert, B Clement, B Cooper, T Delemontex, et al. b-Jet Tagging for Top Physics: Perfomance Studies, Calibrations and Heavy Flavor Fractions. Technical Report ATL-COM-PHYS-2011-124, CERN, Geneva, Feb 2011.

[91] Calibrating the b-Tag Efficiency and Mistag Rate in $35\,\text{pb}^{-1}$ of Data with the ATLAS Detector. Technical Report ATLAS-CONF-2011-089, CERN, Geneva, Jun 2011.

[92] Electron and Photon Reconstruction and Identification in ATLAS: Expected Performance at High Energy and Results at 900 GeV. Technical Report ATLAS-CONF-2010-005, CERN, Geneva, Jun 2010.

[93] The ATLAS Collaboration. Expected Photon Performance in the ATLAS Experiment. Technical Report ATL-PHYS-PUB-2011-007, CERN, Geneva, Apr 2011.

[94] H Abreu, B Brelier, V Dao, M Delmastro, et al. Expected Photon Performance in the ATLAS Experiment. Technical Report ATL-PHYS-INT-2010-137, CERN, Geneva, Dec 2010.

[95] H Abreu, B Brelier, V Dao, M Delmastro, et al. Expected Photon Performance in the ATLAS Experiment. Technical Report ATL-COM-PHYS-2010-240, CERN, Geneva, May 2010.

[96] H. Abreu et al. Measurement of Isolated Di-Photon Cross Section in pp Collision at $\sqrt{s} = 7\,\text{TeV}$ with the ATLAS Detector. Technical Report ATL-COM-PHYS-2011-301, CERN, Geneva, 2011.

[97] Reconstruction and Calibration of Missing Transverse Energy and Performance in Z and W Events in ATLAS Proton-Proton Collisions at 7 TeV. Technical Report ATLAS-CONF-2011-080, CERN, Geneva, Jun 2011.

[98] M Hance, D Olivito, and H Williams. Performance Studies for e/gamma Calorimeter Isolation. Technical Report ATL-COM-PHYS-2011-1186, CERN, Geneva, Sep 2011.

[99] The ATLAS Collaboration. Event Filters.
https://twiki.cern.ch/twiki/bin/viewauth/AtlasProtected/TopPhysD2PDEventFilters .

[100] The ATLAS Collaboration. Luminosity Public Results.
https://twiki.cern.ch/twiki/bin/view/AtlasPublic/LuminosityPublicResults .

[101] The ATLAS Collaboration. COMA Period Documentation.
https://atlas-tagservices.cern.ch/tagservices/RunBrowser .

[102] The ATLAS Collaboration. ATLAS Run Query. http://atlas-runquery.cern.ch.

[103] The ATLAS Collaboration. Top Common Objects.
https://twiki.cern.ch/twiki/bin/viewauth/AtlasProtected/TopCommonObjects2011rel16 ,
2011.

[104] J Hartert and I Ludwig. Electron Isolation in the ATLAS Experiment. Technical Report ATL-COM-PHYS-2010-070, CERN, Geneva, Feb 2010.

[105] M Tripiana and T Dova. Isolation of Photons Revisited. Technical Report ATL-PHYS-INT-2010-014, CERN, Geneva, Jan 2010.

[106] Tracking Results and Comparison to Monte Carlo Simulation at $\sqrt{s} = 900\,\text{GeV}$. Technical Report ATLAS-CONF-2010-011, CERN, Geneva, Jul 2010.

[107] J. Erdmann. Private Communication, 2011.

[108] M. Hance L. Carminati, M. Delmastro et al. Reconstruction and Identification Efficiency of Inclusive Isolated Photons. Technical Report ATL-COM-PHYS-2010-803, CERN, Geneva, Oct 2010.

[109] M. Agustoni et al. Electron Energy Scale in-situ Calibration and Performance. ATL-COM-PHYS-2011-263 (2011).

[110] Georges Aad et al. Electron Performance Measurements with the ATLAS Detector Using the 2010 LHC Proton-Proton Collision Data. *Eur.Phys.J.*, C72:1909, 2012, arXiv:1110.3174.

[111] The ATLAS Collaboration. Energy Scale/Resolution Recommendations.
https://twiki.cern.ch/twiki/bin/viewauth/AtlasProtected/...
...EnergyScaleResolutionRecommendations .

[112] M Baak, M Petteni, and N Makovec. Data-Quality Requirements and Event Cleaning for Jets and Missing Transverse Energy Reconstruction with the ATLAS Detector in Proton-Proton Collisions at a Center-of-Mass Energy of $\sqrt{s} = 7\,\text{TeV}$. Technical Report ATLAS-COM-CONF-2010-038, CERN, Geneva, May 2010.

[113] J F Arguin, D Boumediene, M Bondioli, et al. Jet Selection for Top Physics. Technical Report ATL-COM-PHYS-2010-835, CERN, Geneva, Oct 2010.

[114] The ATLAS Collaboration. How to Clean Jets in 2010 Data.
https://twiki.cern.ch/twiki/bin/viewauth/AtlasProtected/HowToCleanJets .

[115] The ATLAS Collaboration. b-Tagging Benchmarks.
https://twiki.cern.ch/twiki/bin/viewauth/AtlasProtected/BTaggingBenchmarks .

[116] The ATLAS Collaboration. TopETmissLiaison EPS.
https://twiki.cern.ch/twiki/bin/viewauth/AtlasProtected/TopETmissLiaison_EPS .

[117] C. Zhu. Private Twiki Page.
https://twiki.cern.ch/twiki/bin/view/Main/ChengguangZhu .

[118] The ATLAS Collaboration. LAr Cleaining and Object Quality.
https://twiki.cern.ch/twiki/bin/viewauth/AtlasProtected/LArCleaningAndObjectQuality
.

[119] The ATLAS Collaboration. Atlas Production Group.
https://twiki.cern.ch/twiki/bin/view/AtlasProtected/AtlasProductionGroup .

[120] S. Agostinelli et al. GEANT4 - A Simulation Toolkit. *Nucl. Instr. and Meth.*, A506:250, 2003.

[121] Kirill Melnikov, Markus Schulze, and Andreas Scharf. QCD Corrections to Top Quark Pair Production in Association with a Photon at Hadron Colliders. *Phys.Rev.*, D83:074013, 2011, arXiv:1102.1967.

[122] A. Scharf. Private Communication, 2011.

[123] Fleck I. and Scharf A. Private Communication, 2012.

[124] P. M. Nadolsky et al. Implications of CTEQ Global Analysis for Collider Observables. *Phys. Rev.*, D78:013004, 2008, arXiv:0802.0007.

[125] First Tuning of HERWIG/JIMMY to ATLAS Data. Technical Report ATL-PHYS-PUB-2010-014, CERN, Geneva, Oct 2010.

[126] M. et al. Aliev. HATHOR: HAdronic Top and Heavy quarks crOss section calculatoR. *Comput. Phys. Commun.*, 182:1034–1046, 2011, arXiv:1007.1327.

[127] Stefano Frixione, Eric Laenen, Patrick Motylinski, Bryan R. Webber, and Chris D. White. Single-Top Hadroproduction in Association with a W Boson. *JHEP*, 0807:029, 2008, arXiv:0805.3067.

[128] Nikolaos Kidonakis. Next-to-Next-to-Leading-Order Collinear and Soft Gluon Corrections for t-Channel Single Top Quark Production. *Phys.Rev.*, D83:091503, 2011, 1103.2792.

[129] Nikolaos Kidonakis. NNLL Resummation for s-Channel Single Top Quark Production. *Phys.Rev.*, D81:054028, 2010, 1001.5034.

[130] Nikolaos Kidonakis. Two-Loop Soft Anomalous Dimensions for Single Top Quark Associated Production with a W^- or H^-. *Phys.Rev.*, D82:054018, 2010, 1005.4451.

[131] J. Pumplin, D.R. Stump, J. Huston, H.L. Lai, Pavel M. Nadolsky, et al. New Generation of Parton Distributions with Uncertainties from Global QCD Analysis. *JHEP*, 0207:012, 2002, hep-ph/0201195.

[132] S Allwood-Spires, M Barisonzi, H Beauchemin, R Bruneliere, J Buchanan, N Castro, et al. Monte Carlo Samples Used for Top Physics. Technical Report ATL-PHYS-INT-2010-132, CERN, Geneva, Dec 2010.

[133] John M. Campbell and R.K. Ellis. MCFM for the Tevatron and the LHC. *Nucl.Phys.Proc.Suppl.*, 205-206:10–15, 2010, 1007.3492.

[134] The ATLAS Collaboration. Pileup Reweighting.
https://twiki.cern.ch/twiki/bin/view/AtlasProtected/PileupReweighting .

[135] Georges Aad et al. Electron Performance Measurements with the ATLAS Detector using the 2010 LHC Proton-Proton Collision Data. *Eur.Phys.J.*, C72:1909, 2012, arXiv:1110.3174.

[136] The ATLAS Collaboration. Electron Efficiency Measurements.
https://twiki.cern.ch/twiki/bin/viewauth/AtlasProtected/EfficiencyMeasurements ,
2012.

[137] Muon Reconstruction Efficiency in Reprocessed 2010 LHC Proton-Proton Collision Data Recorded with the ATLAS Detector. Technical Report ATLAS-CONF-2011-063, CERN, Geneva, Apr 2011.

[138] Frits A. Berends, W.T. Giele, H. Kuijf, R. Kleiss, and W. James Stirling. Multi-Jet Production in W, Z Events at $p\bar{p}$ Colliders. *Phys.Lett.*, B224:237, 1989.

[139] S.D. Ellis, R. Kleiss, and W. James Stirling. W's, Z's and Jets. *Phys.Lett.*, B154:435, 1985.

[140] A. Caldwell, D. Kollár, and K. Kröninger. BAT - The Bayesian Analysis Toolkit. *Computer Physics Communications*, 180:2197–2209, November 2009, arXiv:0808.2552.

[141] R Daya, R Ishmukhametov, D Joffe, and R Stroynowski. Measurement of Electron-Photon Fake Rate in ATLAS Official Monte Carlo Data Using $Z \to ee$ Decays. Technical Report ATL-PHYS-INT-2010-085, CERN, Geneva, Sep 2010.

[142] K Becker, A Cortes Gonzalez, V Dao, et al. Mis-Identified Lepton Backgrounds in Top Quark Pair Production Studies for EPS 2011 Analyses. Technical Report ATL-COM-PHYS-2011-768, CERN, Geneva, Jun 2011.

[143] The ATLAS Collaboration. MCTruthClassifier.
https://twiki.cern.ch/twiki/bin/viewauth/AtlasProtected/MCTruthClassifier .

[144] K Perez, G Romeo, and A Schwartzman. Jet Energy Resolution and Reconstruction Efficiencies from in-situ Techniques with the ATLAS Detector Using Proton-Proton Collisions at a Center of Mass Energy $\sqrt{s} = 7\,\text{TeV}$. Technical Report ATLAS-COM-CONF-2010-056, CERN, Geneva, Jun 2010.

[145] Hung-Liang Lai, Marco Guzzi, Joey Huston, Zhao Li, Pavel M. Nadolsky, et al. New Parton Distributions for Collider Physics. *Phys.Rev.*, D82:074024, 2010, arXiv:1007.2241.

[146] G. Watt and R.S. Thorne. Study of Monte Carlo Approach to Experimental Uncertainty Propagation with MSTW 2008 PDFs. *JHEP*, 1208:052, 2012, arXiv:1205.4024.

[147] Richard D. Ball, Luigi Del Debbio, Stefano Forte, Alberto Guffanti, Jose I. Latorre, et al. A First Unbiased Global NLO Determination of Parton Distributions and their Uncertainties. *Nucl.Phys.*, B838:136–206, 2010, arXiv:1002.4407.

[148] The ATLAS Collaboration. CERN SVN: MC10 Job Options.
https://svnweb.cern.ch/trac/atlasoff/browser/Generators/MC10JobOptions .

[149] Luminosity Determination in pp Collisions at $\sqrt{s} = 7\,\text{TeV}$ using the ATLAS Detector in 2011. Technical Report ATLAS-CONF-2011-116, CERN, Geneva, Aug 2011.

i want morebooks!

Buy your books fast and straightforward online - at one of world's fastest growing online book stores! Environmentally sound due to Print-on-Demand technologies.

Buy your books online at
www.get-morebooks.com

Kaufen Sie Ihre Bücher schnell und unkompliziert online – auf einer der am schnellsten wachsenden Buchhandelsplattformen weltweit! Dank Print-On-Demand umwelt- und ressourcenschonend produziert.

Bücher schneller online kaufen
www.morebooks.de

 VDM Verlagsservicegesellschaft mbH
Heinrich-Böcking-Str. 6-8   Telefon: +49 681 3720 174   info@vdm-vsg.de
D - 66121 Saarbrücken   Telefax: +49 681 3720 1749   www.vdm-vsg.de

Printed by Books on Demand GmbH, Norderstedt / Germany